Practitioner's Knowledge Representation

Emilia Mendes

Practitioner's Knowledge Representation

A Pathway to Improve Software Effort Estimation

 Springer

Emilia Mendes
Blekinge Institute of Technology
Karlskrona
Sweden

ISBN 978-3-662-51199-2   ISBN 978-3-642-54157-5 (eBook)
DOI 10.1007/978-3-642-54157-5
Springer Heidelberg New York Dordrecht London

Printed on acid-free paper

Springer is part of Springer Science+Business Media (www.springer.com)

This book is dedicated to the wonderful woman who gave me life, strength and love... my mother

Preface

The main goal in writing this book was to help organisations improve their effort estimates and their effort estimation processes by providing a step-by-step methodology that takes them through the building and validation of models that are based on their own knowledge and experience. Such models, once validated, can be used to obtain predictions, carry out risk analyses, help organisations with their decision-making when estimating effort for new projects and set a pathway to making those organisations into learning organisations.

This methodology, called expert-based knowledge engineering of Bayesian networks (EKEBNs), has been adapted by the author as a result of several collaborations with six different companies in New Zealand and Brazil. Domain experts from each company participated in the elicitation of bespoke models for effort estimation. The building of such models led those companies to change their estimation processes and to also improve their estimates. Their stories are detailed in Chaps. 7–12. Note that the methodology detailed in this book can also be employed to build models aiming at different goals other than effort estimation (e.g., quality prediction, risk management, resource management and prediction).

All models were built using a single tool, called Netica. This tool was chosen since it provided simplicity and the functionality that was needed to carry out the work. The example model that is used in some parts of this book was also created using this same tool. This model is available for download, and the tool is also free to use with models that do not contain more than 15 factors. We hope that making the example model available will encourage companies to run the model and see the value in using such models for decision-making.

Writing this book was made possible due to the participation of several companies in New Zealand and Brazil with which I had the privilege to collaborate and research funding from the Royal Society of New Zealand and from the Brazilian Government (CAPES/PVE).

Karlskrona, Sweden   Emilia Mendes
October 2013

Contents

Introduction to Knowledge Management

1

Introduction

At the heart of an organisation's ability to sustain its competitive advantage and to innovate are the knowledge it holds and its capability to learn and utilise such knowledge [1, 2]. This idea is supported by a growing number of publications in areas such as knowledge management, knowledge-creating companies and learning organisation (e.g., [3–5]). Garvin defines a learning organisation as follows [2]:

> "A learning organisation is an organisation skilled at creating, acquiring, and transferring knowledge, and at modifying its behaviour to reflect new knowledge and insights".

Clearly, sustainable organisational improvement requires a "commitment to learning" [6, 7].

If we consider software organisations, regardless of whether they manage Web-based or software projects, the core of what they do is knowledge intensive [4, 7, 8]. However, their use of knowledge management activities is often still lacking, and is far from changing them into learning organisations [8]. A recent systematic literature review on knowledge management in software engineering presented the following gaps in this area [5]:

1. Software engineering has predominantly only addressed the storage and retrieval of knowledge, and has ignored other important aspects such as knowledge creation, transfer and application;
2. There has been no identification to date of success factors for knowledge management in software engineering;
3. There is a siloed view of knowledge management by organisations as they tend to use solely tacit or explicit knowledge, rather than to combine both as part of a continuous process;
4. There has been a focus of agile software development mainly on tacit knowledge-driven management activities, and a focus of traditional software development mainly on explicit knowledge-driven management activities.

E. Mendes, *Practitioner's Knowledge Representation*, DOI 10.1007/978-3-642-54157-5_1,   1
© Springer-Verlag Berlin Heidelberg 2014

Knowledge management and decision making are intrinsically related, given that the quality of the decisions taken is extremely likely to be influenced by the usefulness and effective representation of the knowledge used in those decisions. Whenever decisions are carried out within the scope of a complex knowledge domain (e.g., software project/product management), they present an uncertain nature. Note that herein uncertain means that the knowledge is based on beliefs and therefore cannot be assumed to be absolute with deterministic outcomes [9].

The literature in the field of decision-making advocates that a suitable solution to support decision-making under uncertainty is to build models that make explicit decision makers' mental models [10], as such models can be used to compare different decision scenarios and hence provide better understanding of the situation at hand [10, 11]. This means that any knowledge management activities being employed by an organisation must include the building of explicit models representing experts' mental models (tacit knowledge). In addition, decisions (how one sees, thinks or acts in the world) are influenced by decision makers' mental models [10]; therefore, updating and enriching these mental models leads to improved decision-making processes [6, 10]. Mental models (a.k.a. representations and cognitive maps [11]) are enhanced through the use of a knowledge creation process [2, 6, 12], which is discussed next.

Knowledge creation is one of the three different processes (knowledge creation, transfer and application) embedded into the theory of organisational knowledge creation by Nonaka and Toyama [6]. This theory is the basis for all the improvement actions that are detailed throughout this book, and is introduced next.

Nonaka and Toyama's theory is cited in numerous knowledge management studies (e.g., [1, 2, 7]), and has also been used to guide improvement activities in software process improvement studies, with extremely promising results [3, 4]. It organises the knowledge process into four different stages [6] (see Fig. 1.1):

1. Tacit to tacit, where experiences, skills and expertise are shared between individuals. A typical example is that of a carpenter teaching carpentry to a helper. The carpenter explains and shows how to accomplish the tasks; however, all the knowledge transferring that takes place is done via socialisation only, without any written manuals or guides. The end result of such type of sharing is a change in the helper's mental model due to the tacit knowledge transferring that took place. In other words, when knowledge is transferred on a tacit to tacit level the end result is not a concrete, tangible representation of that knowledge. Another example, however, now in the area of software engineering, is the following: two senior project managers explain verbally to a junior project manager the approach they use when estimating the effort (person-hours) for a new project. This approach is based on expert knowledge only. In a similar way to the previous example, this scenario also describes a situation in which knowledge is transferred but does not result in a tangible knowledge representation (e.g., a sketch, a manual) that details the tacit knowledge that is used (and how it is used) in order to achieve a particular goal. Both examples illustrate knowledge being transmitted solely via discussions (socialisation), which leads

Fig. 1.1 The four different stages of the theory of organisational knowledge creation (Image prepared by Ms. Jacy Rabelo)

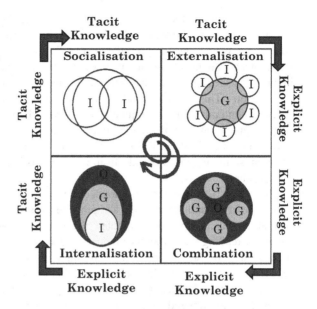

to learned skills and shared mental models. In other words, knowledge is being transmitted at the tacit level only, without any documentation/model/tangible representation making explicit the tacit knowledge that the carpenter and the two senior project managers have relating to their experiences and knowledge in carpentry and project management, respectively;

2. Tacit to explicit, where tacit knowledge is "translated" by an individual or by a group into an explicit (tangible) representation. If we revisit the two examples presented in (1), they could read as follows: a carpenter is writing a manual documenting how to build several furniture items (e.g., chair, table) using explanations and examples in natural language; the second example could be as follows: two senior project managers are working together preparing a detailed set of training material (slides) and guidelines (booklet), all written in natural language, to be used to explain to junior project managers the approach those senior managers employ when estimating effort for new projects. Both scenarios provide examples of some of the possible ways in which tacit knowledge can be made explicit. Note that in both scenarios the end result is a tangible representation of tacit knowledge—a manual, slides and a booklet. These tangible representations all use natural language as a way to explicitly characterise experts' tacit knowledge; however, the choice of whether to use natural language or an alternative choice for knowledge representation should be driven by the main goal that motivated the explicitation of tacit knowledge. Note that the explicitation of tacit knowledge has numerous advantages, including wider sharing of existing expert knowledge within an organisation;

3. Explicit to explicit, where explicit knowledge from different groups is gathered, combined, edited and diffused. An example is to combine carpentry manuals written by different experienced carpenters/groups of carpenters, and also to

combine training material and guidelines previously prepared by senior project managers. Here all the knowledge being combined represents tangible knowledge that was previously represented explicitly from tacit knowledge. There are numerous advantages to an organisation in combing explicit knowledge, which include the creation of a common understanding about a particular practice/process, etc. This common understanding can be used as the basis for best-practice standards to be disseminated and followed throughout the organisation. Such aggregation of knowledge should also be complemented with explicit knowledge of the state of the art and the state of the practice, thereby enabling an organisation to learn not only from its own existing expertise, but also from that of others.

4. Explicit to tacit, where explicit knowledge is absorbed by individuals in groups within the organisation via action and practice, thus enhancing those individuals' mental models. Using the examples already given, we could have a junior carpenter reading the carpentry manuals and applying what is prescribed in the manual step by step, and as a result internalising the tacit knowledge via experiencing the process. In terms of our other example, we could have a junior project manager who has just participated in project-management training simulate a scenario prepared by a senior project manager, in which effort has to be estimated for a new project.

Knowledge creation is meant to be a continuous process that, as an integral part, traverses all four stages, i.e., to be a knowledge spiral.

Once the explicitation of a mental model (or several mental models combined) has taken place via the use of a knowledge creation process, this explicitated model can be employed as part of a decision-making process. Let's look at an example that brings together the concepts of knowledge management and decision making.

In this example we consider a consulting company that provides Web-based solutions to clients. This company has three project managers who are responsible for all the effort estimates prepared for their new projects. These three project managers attended a seminar by a researcher who has collaborated with other companies in order to help them improve their decision making relating to effort estimation. As a consequence of attending the seminar, these three project managers decided to collaborate with the researcher (called here a knowledge engineer, KE) in order to improve their decision making relating to the effort estimated for each of their new projects. This collaboration took place via weekly meetings, which were attended by the three project managers and the KE and where a knowledge creation process was employed, led to the model presented in Fig. 1.2.

This model, which is a tangible representation of the combined mental models from the three project managers based on their experience managing Web projects and estimating effort, shows what these managers believe to be the fundamental factors affecting effort estimates, how these factors are inter-related, and also their quantification of the uncertainty that is inherent to the domain of effort estimation (all the bars and numbers listed with each of the factors). We refer to this as an example model, EM, henceforth. Note that a step-by-step description of

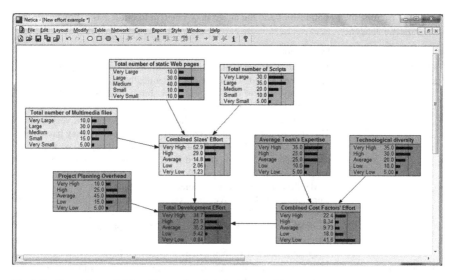

Fig. 1.2 Example model for effort estimation decision making

the type of model that is presented herein and the process used to build such tangible model are provided in Chaps. 5 and 6, respectively. The factors that are included in the EM are detailed in Table 1.1.

Figure 1.2 shows that, based upon the three projects managers' past experience, 40 % of the past Web projects had a medium number of static Web pages, 35 % had a large number of scripts and 40 % had a medium number of multimedia files. It also shows that the effort of the combined sizes was very high 52.9 % of the time, and so on.

How can such a model be used for decision making?

Let's consider three different decision-making scenarios, and also assume that the categories that are chosen for some of the Factors are based on the results from requirements elicitation meetings with clients.

Scenario 1: After attending a few requirements elicitation meetings with a client, one of the project managers has estimated categories for six different Factors, as follows:

Factor	Category selected
Total number of static Web pages	Medium
Total number of scripts	Very large
Total number of multimedia files	Medium
Average team's expertise	Average
Technological diversity	High
Project planning overhead	High

Table 1.1 Factors, their relationships and the uncertainty inherent to effort estimation as per the three project managers

Factor	Description
Total number of static Web pages	This factor represents the estimated number of new Web pages that need to be created. These are not dynamically-generated pages, and include any type of page such as .htm, .html, .php. Figure 1.2 also shows that the total number of static Web pages is measured using five different categories (very large, large, medium, small and very small). This means that when project managers use the model, they have these five categories to choose from, and the choice will depend on the set of requirements they have gathered from the client for whom this application is to be developed. Within the context of the example model (EM), these categories are detailed as follows: Very small number of Web pages → 1–5 Web pages; Small number of Web pages → 6–15 Web pages; Medium number of Web pages → 16–25 Web pages; Large number of Web pages → 26–30 Web pages; Very large number of Web pages → 31+ Web pages
Total number of scripts	This factor represents the estimated sum of any types of scripts that are likely to be created for the Web application. Such scripts can be made using a client-side scripting technique (e.g., XML, Ajax techniques, Flash ActionScript), or server-side scripting languages (ASP, JSP, Perl, PHP, Python). It also includes files written using cascading style sheets (css). Figure 1.2 also shows that the total number of scripts is measured using five different categories (very large, large, medium, small and very small). Within the context of the EM, these categories are detailed as follows: Very small number of scripts/css files → 0–7 scripts/css files; Small number of scripts/css files → 8–20 scripts/css files; Medium number of scripts/css files → 21–25 scripts/css files; Large number of scripts/css files → 26–35 scripts/css files; Very large number of scripts/css files → 36+ scripts/css files
Total number of multimedia files	This factor represents the total estimated number of any multimedia content, such as images and videos. Figure 1.2 shows that the total number of multimedia files is measured using five different categories (very large, large, medium, small and very small). Within the context of the EM, these categories are detailed as follows: Very small number of multimedia content → 0–3 multimedia content; Small number of multimedia content → 4–8 multimedia content; Medium number of multimedia content → 9–20 multimedia content; Large number of multimedia content → 21–30 multimedia content; Very large number of multimedia content → 31+ multimedia content
Combined size's effort	This factor represents the estimated amount of effort (person-hours) needed to create Web pages, scripts/css files and multimedia files. Note that the effort changes depending on which categories are selected for each factor. Such selection takes place as part of a decision-making scenario, and examples are given later on (see Figs. 1.3, 1.4, and 1.5). Figure 1.2 shows that combined size's effort is measured using five different categories (very high, high, average, low and very low). Within the context of the EM, these categories are detailed as follows: Very low effort → 1–40 person-hours; Low effort → 40+ to 80 person-hours

(continued)

Table 1.1 (continued)

Factor	Description
	Average effort → 80+ to 160 person-hours High effort → 160+ to 320 person-hours Very high effort → 320+ person-hours
Average team's expertise	This factor measures team expertise as the average number of years of experience that the development team has with Web development. The estimation within this context relates to tentative decision as to who will likely be allocated to the team that will develop the Web application for which total effort is being estimated Figure 1.2 shows that five different categories (very high, high, average, low and very low) are used to measure the average team's expertise. Within the context of the EM, these categories are detailed as follows: Very low team's expertise → 1 year of experience Low team's expertise → 2–3 years of experience Average team's expertise → 4–8 years of experience High team's expertise → 9–12 years of experience Very high team's expertise → 13+ years of experience
Technological diversity	This factor represents the estimated amount of diversity as far as the use of technology is concerned. It is measured using a surrogate measure, represented by the number of different technologies that are being employed in order to develop a Web application. Examples of technologies are MySQL, PHP, HTML, CSS, Python, ASP and JSP. Five categories are employed to measure technological diversity. Within the context of the EM, these categories are detailed as follows: Very low technological diversity → 1 type of technology is being used in the Web application Low technological diversity → 2 different types of technology are being used in the Web application Average technological diversity → 3–4 different types of technology are being used in the Web application High technological diversity → 5–7 different types of technology are being used in the Web application Very high technological diversity → 8+ different types of technology are being used in the Web application
Combined cost factors' effort	This factor represents the estimated amount of effort (person-hours) when taking into account technological diversity and average team's expertise. Note that the effort changes depending on which categories are selected for each factor. Such selection takes place as part of a decision-making scenario, and examples are given later on (see Figs. 1.3, 1.4 and 1.5). Figure 1.2 shows that the combined cost factors' effort is measured using five different categories (very high, high, average, low and very low). Within the context of the EM, these categories are detailed as follows: Very low effort → 1–80 person-hours Low effort → 80+ to 200 person-hours Average effort → 200+ to 400 person-hours High effort → 400+ to 800 person-hours Very high effort → 800+ person-hours
Project planning overhead	This factor represents the degree of participation needed by the project manager in order to ensure the project is managed adequately and is ideally completed within time and on budget. This includes, but is not

(continued)

Table 1.1 (continued)

Factor	Description
	limited to, status reports; communication; implementation plan (for large projects), which includes the tasks to be done and their estimated completion dates; risk analysis; data analysis; planning (project execution plan) Figure 1.2 shows that the project planning overhead is measured in our EM using five different categories (very high, high, average, low and very low), which are detailed as follows: Very low project overhead → 5 % of estimated effort Low project overhead → 15 % of estimated effort Average project overhead → 20 % of estimated effort High project overhead → 30 % of estimated effort Very high project overhead → 40 % of estimated effort
Total development effort	This factor represents the total estimated effort to develop a Web application. The three factors that have a direct effect upon total effort are: combined cost factors' effort, project planning overhead and combined size's effort. Figure 1.2 shows that the total development effort is also measured in our EM using five different categories (very high, high, average, low and very low), which are detailed as follows: Very low effort → 1–126 person-hours Low effort → 126+ to 320 person-hours Average effort → 320+ to 670 person-hours High effort → 670+ to 1,400 person-hours Very high effort → 1,400+ person-hours

Such choices are also shown in Fig. 1.3.

Once these categories were selected, the model suggested that the total estimated effort was 63.6 % likely to be very high. However, there was also a 31.5 % chance that total effort could be high and a 4.92 % chance that it could be average. However, despite the strong suggestion by the model that total effort was very likely to be very high, the project manager decided to try a different scenario (Scenario 2) as he was not sure whether the team available to work on the project would have average expertise.

Scenario 2: Except for the factor "Average Team's Expertise" this scenario is mostly the same as Scenario 1. Here the EM also suggests that the total estimated effort is very likely to be very high; however, now with a certainty of 76 %, as opposed to 63.6 % (see Fig. 1.4).

Factor	Category selected
Total number of static Web pages	Medium
Total number of scripts	Very large
Total number of multimedia files	Medium
Average team's expertise	Very low
Technological diversity	High
Project planning overhead	High

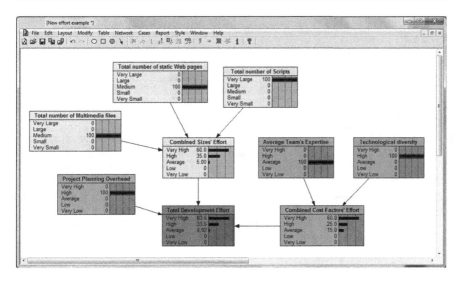

Fig. 1.3 Example model for scenario 1

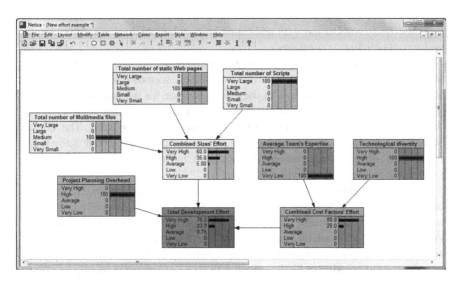

Fig. 1.4 Example model for scenario 2

Both scenarios strongly suggest that total effort is likely to be very high. Such scenarios, also known as what-if scenarios, provide the means to estimate effort using as basis the knowledge and experience from the project managers; therefore, such scenarios can provide support for decisions relating to effort estimates for new projects. Both scenarios show the use of the EM for predictive reasoning, i.e., to make predictions using as basis the existing knowledge already embedded in the

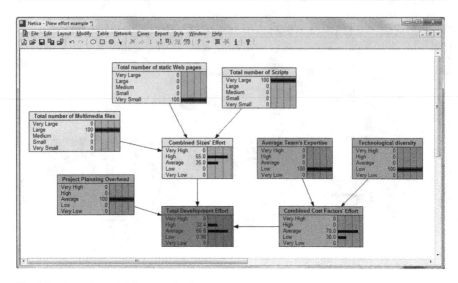

Fig. 1.5 Example model for scenario 3

model itself, in combination with additional knowledge obtained via, for example, requirements elicitation meetings.

Once several what-if scenarios are run and compared, the project manager can use the suggested effort estimate from one of these scenarios (or a combination), without any further modification, as the effort estimate for the new project, or she can also compare the output from the model with her own independently prepared subjective effort estimate (at least for a number of new projects). The choice between relying solely on the model or to also use experts' subjective knowledge for some time depends on the amount of validation that took place when building the model. Further details are provided in Chap. 6.

There is no limitation on the number of scenarios that can be created (see Scenario 3 and Fig. 1.5), and they can also be readily used in discussions between the project manager(s) and the prospective development team(s). In addition, such models can also be used by junior project managers, thereby enabling them to understand the factors that were previously chosen by more experienced managers and to also run scenarios to be discussed with more experienced managers. These activities will lead to tacit knowledge learning, which is also an important aspect of a learning organisation.

Scenario 3: This scenario represents the following choice of categories:

Factor	Category selected
Total number of static Web pages	Very small
Total number of scripts	Very large
Total number of multimedia files	Large
Average team's expertise	Low
Technological diversity	Low
Project planning overhead	Average

The EM and the different scenarios presented in this chapter illustrate the use of a knowledge creation process where tacit knowledge is explicitated and represented as a model that can then be used via what-if scenarios to support decision making and tacit knowledge learning.

The core of this book is to explain how such models can be created via the elicitation of tacit knowledge from domain experts. There are 14 chapters in this book, which are briefly introduced below:

Chapter 1 provides an introduction to knowledge management, how its principles can be used when building explicit mental models and how such models can be used for decision making.

Chapter 2 provides a discussion relating to differences between Web and software applications and development processes. This is done in order to provide a clear motivation for why effort estimation within the scope of Web development is needed and how Web applications differ from traditional software applications.

Chapter 3 provides an introduction to effort estimation to provide all readers with an understanding of this fundamental process that is part of any Web and software project management. We also include this chapter in the book in order to improve the understanding of the case studies detailed in Chaps. 7–12, which are aimed at building effort estimation models.

Chapter 4 provides an overview of the state of the art in the area of Web effort estimation, based on the findings from a recent systematic literature review on Web resource estimation.

Chapter 5 provides an introduction to Bayesian networks, which are employed in all the case studies discussed in this book.

Chapter 6 details the step-by-step methodology that was used in all six case studies on building effort estimation models, presented in Chaps. 7–12.

Chapters 7–12 each present a separate case study where effort estimation models were built in collaboration with domain experts from different companies in Auckland, New Zealand and Rio de Janeiro, Brazil.

Chapter 13 elaborates on the many different ways a company can use models such as those that are the focus of this book. It also discusses the issue of aggregating different models and provides a suggestion on how to aggregate.

Chapter 14 provides conclusions to the book and summarises the main messages that were presented throughout the previous 13 chapters.

Conclusions

This chapter introduced the principles behind knowledge management and knowledge organisation, and discussed how such principles can be incorporated into explicit expert-based models (mental models). We stressed in particular that knowledge creation is an iterative process as proposed by Nonaka and Toyama [6]. It provides a pathway for building expert-based mental models, which can also be used for decision making at the group, division, department and even organisation level. We ended the chapter with an overview of the remaining chapters of this book.

References

1. Dyba T (2003) A dynamic model for software engineering knowledge creation. In: Aurum A, Jeffery R, Wohlin C, Handzic M (eds) Managing software engineering knowledge. Springer, Berlin, pp 95–114
2. Garvin DA (1998) Building a learning organization. In: Harvard business review on knowledge management. Harvard Business Publishing, Boston, MA
3. Arent J, Nørbjerg J (2000) Software process improvement as organizational knowledge creation: a multiple case analysis. In: Proceedings of the 33rd HICSS conference, pp 1–11
4. Aurum A, Jeffery R, Wohlin C, Handzic M (eds) (2003) Managing software engineering knowledge. Springer, Berlin
5. Bjørson FO, Dingsøyr T (2008) Knowledge management in software engineering: a systematic review of studied concepts, findings and research methods used. Inf Softw Technol 50(11): 1055–1068
6. Nonaka I, Toyama R (2003) The knowledge-creating theory revisited: knowledge creation as a synthesizing process. Knowl Manag Res Pract 1:2–10
7. Schneider K (2009) Experience and knowledge management in software engineering. Springer, Berlin
8. Dingsøyr T, Bjørson FO, Shull F (2009) What do we know about knowledge management? Practical implications for software engineering. IEEE Softw 26(3):100–103
9. Pearl J (1988) Probabilistic reasoning in intelligent systems: networks of plausible inference. Morgan Kaufmann, San Francisco, CA
10. Steiger DM (2010) Decision support as knowledge creation: a business intelligence design theory. Int J Bus Intell Res 1:29–47
11. Chermack TJ (2003) Mental models in decision making and implications for human resource development. Adv Dev Hum Resour 5:408–422
12. Lempert R, Nakicenovic N, Sarewitz D, Schlesinger M (2004) Characterizing climate-change uncertainties for decision-makers. An editorial essay. Clim Chang 65:1–9

Web Development Versus Software Development

2

Introduction

The World Wide Web (Web) was originally conceived in 1989 as an environment to allow for the sharing of information (e.g., research reports, databases, user manuals) amongst geographically dispersed individuals. The information itself was stored on different servers and was retrieved by means of a single user interface (Web browser). The information consisted primarily of text documents inter-linked using a hypertext metaphor (http://www.zeltser.com/web-history/) [1].

Since its original inception, the Web has changed into an environment employed for the delivery of many different types of applications. Such applications range from small-scale information-dissemination-like applications, typically developed by writers and artists, to large-scale commercial, enterprise-planning and scheduling, collaborative-work applications. The latter are developed by multidisciplinary teams of people with diverse skills and backgrounds using cutting-edge, diverse technologies [1–3]. The increase in the use of the Web to provide commercial applications has been motivated by several factors, such as the possible increase of an organisation's competitive position, and the opportunity for small organisations to project their corporate presence in the same way as that of larger organisations [4]. Numerous current Web applications are fully functional systems that provide business-to-customer and business-to-business e-commerce, and numerous services to numerous users [1].

Industries such as travel and hospitality, manufacturing, banking, education and government utilised Web-based applications to improve and increase their operations [3]. In addition, the Web allows for the development of corporate intranet Web applications, for use within the boundaries of individual organisations [5]. The remarkable spread of Web applications into areas of communication and commerce makes it one of the leading and most important branches of the software industry [1].

Web development is a relatively new and rapidly growing industry, with e-commerce alone weathering the recession and growing 11 % in the United States

E. Mendes, *Practitioner's Knowledge Representation*, DOI 10.1007/978-3-642-54157-5_2,   13
© Springer-Verlag Berlin Heidelberg 2014

in 2009, with similar growth in 2010.[1] This continued growth makes it worthwhile to conduct research that enables Web development companies to make more efficient managerial decisions [6].

To date, the development of Web applications has generally been ad hoc, resulting in poor-quality applications, which are difficult to maintain [1]. The main reasons for such problems are unsuitable design and development processes, and poor project management practices [3]. A survey on Web-based projects, published by the Cutter Consortium in 2000, revealed a number of problems with outsourced large Web-based projects [3]:

- 84 % of surveyed delivered projects did not meet business needs.
- 53 % of surveyed delivered projects did not provide the required functionality.
- 79 % of surveyed projects presented schedule delays.
- 63 % of surveyed projects exceeded their budget.

As the reliance on larger and more complex Web applications increases, so does the need for using methodologies and best practice guidelines to develop applications that are delivered on time, within budget, have a high level of quality and are easy to maintain [7–9]. To develop such applications Web development teams need to use sound methodologies, systematic techniques, quality assurance, rigorous, disciplined and repeatable processes, better tools, and baselines. Web engineering[2] aims to meet such needs [11].

Web engineering is described as [12]:

> the use of scientific, engineering, and management principles and systematic approaches with the aim of successfully developing, deploying and maintaining high quality Web-based systems and applications.

This is a similar definition to that used to describe software engineering; however, both disciplines differ in many ways. Such differences are discussed next.

Web Applications Versus Conventional Software

An overview of differences between Web and software development with respect to their development processes, technologies, quality factors, and measures is presented here. In addition, this section also provides definitions and terms used throughout the book (e.g., Web application).

---

[1] http://blogs.wsj.com/digits/2010/03/08/e-commerce-growth-slows-but-still-out-paces-retail/

[2] The term "Web engineering" was first published in 1996 in a conference paper by Gellersen et al. [10]. Since then this term has been cited in numerous publications, and numerous activities devoted to discussing Web engineering have taken place (e.g., workshops, conference tracks, entire conferences).

Web Hypermedia, Web Software or Web Application?

The Web is the best-known example of a hypermedia system. To date, numerous organisations world-wide have developed a vast array of commercial and/or educational Web applications. The Web literature uses numerous synonyms for a Web application, such as Web site, Web system, Internet application. The IEEE Std 2001–2002 uses the term Web site defined as [13]:

"A collection of logically connected Web pages managed as a single entity."

However, using Web site and Web application interchangeably does not allow one to differentiate between the physical storage of Web pages and their application domains.

The Web has been used as the delivery platform for three types of applications: Web hypermedia applications, Web software applications, and Web applications [14].

- *Web hypermedia application*—a nonconventional application characterised by the authoring of information using nodes (chunks of information), links (relations between nodes), anchors, access structures (for navigation), and delivery over the Web. Technologies commonly used for developing such applications are HTML, XML, JavaScript and multimedia. In addition, typical developers are writers, artists and organisations who wish to publish information on the Web and/or CD-ROMs without the need to know programming languages such as Java. These applications have unlimited potential in areas such as software engineering, literature, education and training.
- *Web software application*—a conventional software application that relies on the Web or uses the Web's infrastructure for execution. Typical applications include legacy information systems such as databases, booking systems, knowledge bases, etc. Many e-commerce applications fall into this category. Typically they employ development technologies (e.g., DCOM, ActiveX, etc.), database systems, and development solutions (e.g., J2EE). Developers are in general young programmers fresh from a Computer Science or Software Engineering degree course, managed by a few more senior staff.
- *Web application*—an application delivered over the Web that combines characteristics of both Web hypermedia and Web software applications.

Web Development Versus Software Development

Web development and software development differ in a number of areas, which will be detailed later. However, of these, three such areas seem to provide the greatest differences and to affect the entire Web development and maintenance processes. These areas encompass the people involved in development, the intrinsic characteristics of Web applications and the audience for which they are developed.

The development of conventional software remains dominated largely by IT professionals, where a sound knowledge of programming, database design, and project management is necessary. In contrast, Web development encompasses a much wider variety of developers, such as amateurs with no programming skills, graphics designers, writers, database experts and IT professionals, to name but a few. This is possible as Web pages can be created by anyone without the necessity for programming knowledge [15].

Web applications by default use communications technology and have multi-platform accessibility. In addition, since they employ a hypermedia paradigm, they are non-sequential by nature, using hyperlinks to interrelate Web pages and other documents. Therefore, navigation and pluralistic design become important aspects to take into account. Finally, the multitude of technologies available for developing Web applications means that developers can build a full spectrum of applications, from a static simple Web application using HTML to a fully-fledged distributed e-commerce application [7]. Conventional software can be developed using several programming languages running on a specific platform, components off the shelf (COTS), etc. It can also use communications technology to connect to and use a database system. However, the speed of implementing new technology is faster for Web development relative to non-Web-based applications.

Web applications are aimed at wide-ranging groups of users. Such groups may be known ahead of time (e.g., applications available within the boundaries of an intranet). However, it is more often the case that Web applications are devised for an unknown group of users, making the development of aesthetically pleasing applications more challenging [16]. In contrast, conventional software applications are generally developed for a known user group (e.g., department, organisation) making the explicit identification of target users an easier task.

For the purpose of our discussion, we have grouped the differences between Web and software development into 12 areas, which are as follows:

1. Application characteristics
2. Primary technologies used
3. Approach to quality delivered
4. Development process drivers
5. Availability of the application
6. Customers (stakeholders)
7. Update rate (maintenance cycles)
8. People involved in development
9. Architecture and network
10. Disciplines involved
11. Legal, social and ethical issues
12. Information structuring and design

(1) *Application Characteristics*

   Web applications are created by integrating numerous distinct elements, such as fine-grained components (e.g., DCOM, OLE, ActiveX), interpreted scripting languages, components off the shelf (COTS, e.g., customised applications, library components, third-party products), multimedia files (e.g., audio, video, 3D objects), HTML/SGML/XML files, graphical images, mixtures of HTML and programs, and databases [16–18]. Components may be integrated in many different ways and present different quality attributes. In addition, their source code may be proprietary or unavailable, and may reside on and/or be executed from different remote computers [17]. Web applications are, for the large part, platform-independent (although there are exceptions, e.g., OLE, ActiveX), and Web browsers in general provide similar user interfaces with similar functionality, freeing users from having to learn distinct interfaces [16]. Finally, a noticeable difference between Web applications and conventional software applications is in the use of navigational structures. Web applications use a hypermedia paradigm where content is structured and presented using hyperlinks. Navigational structures may also need to be customised, i.e., by the dynamic adaptation of content structure, atomic hypermedia components, and presentation styles [10].

Despite the initial attempt by the hypermedia community to develop conventional applications with a hypermedia-like interface, largely conventional software applications do not employ this technique.

Again in contrast, conventional software applications can also be developed using a wide variety of components (e.g., COTS), generally developed using conventional programming languages such as C++, Visual Basic, and Delphi. These applications may also use multimedia files, graphical images and databases. It is common that user interfaces are customised depending on the hardware, operating system, software in use and the target audience [16]. There are programming languages on the market (e.g., Java) that are intentionally cross-platform; however, the majority of conventional software applications tend to be monolithic, running on a single operating system.

(2) *Primary Technologies Used*

Web applications are developed using a wide range of diverse technologies, such as the many flavoured Java solutions (Java servlets, Enterprise JavaBeans, applets, and JavaServer Pages), HTML, JavaScript, XML, UML, databases and much more. In addition, there is an increasing use of third-party components and middleware. Since Web technology is an area that changes quickly, some authors suggest it may be difficult for developers and organisations to keep up with what is currently available [17].

The primary technology used to develop conventional software applications is mostly represented by object-oriented methods, generators and languages, relational databases, and CASE tools [18]. The pace with which new technologies are proposed is slower than that for Web applications.

(3) *Approach to Quality Delivered*

Web companies that operate their business on the Web rely heavily on providing applications and services of high quality so that customers return to do repeat business. As such, these companies only see a return on investment if customers' needs have been fulfilled. Customers who use the Web for obtaining services have very little loyalty to the companies they do business with. This suggests that new companies providing Web applications of a higher quality will most likely displace customers from previously established businesses. Further, that quality is the principal factor that brings repeated business. For

Web development, quality is often considered a higher priority than time to market, with the mantra "later and better" as the mission statement for Web companies who wish to remain competitive [17].

Within the context of conventional software development, software contractors are often paid for their delivered application regardless of its quality. Return on investment is immediate. Ironically, they are also often paid for fixing defects in the delivered application, where these failures principally exist because the developer did not test the application thoroughly. This has the knock-on effect that a customer may end up paying at least twice (release and fixing defects) the initial bid in order to make the application functional. Here time to market takes priority over quality, since it can be more lucrative to deliver applications with plenty of defects sooner than high-quality applications later. For these companies the "sooner but worse" rule applies [17].

Another popular mechanism employed by software companies is to fix defects and make the updated version into a new release, which is then resold to customers, bringing in additional revenue.

(4) *Development Process Drivers*

The dominant development process drivers for Web companies have three quality criteria [17]:

- reliability,
- usability and
- security,

followed by:

- availability,
- scalability,
- maintainability and
- time to market.

Reliability: applications that work well, do no crash, do not provide incorrect data, etc.

Usability: an application that is simple to use. If a customer wants to use a Web application to buy a product on-line, the application should be as simple to use as the process of physically purchasing that product in a shop. Many existing Web applications present poor usability despite the extensive range of Web usability guidelines that have been published. A Web application with poor usability will quickly be replaced by another more usable application as soon as its existence becomes known to the target audience [17].

Security: the handling of customer data and other information securely so that problems such as financial loss, legal consequences and loss of credibility can be avoided [17].

With regards to conventional software development, the development process driver is time to market and not quality criteria [17].

(5) *Availability of the Application*

Customers who use the Web expect applications to be operational throughout the whole year (24/7/365). Any downtime, no matter how short, can be detrimental [17].

Except for a few application domains (e.g., security, safety critical, military, banking) customers of conventional software applications do not expect these applications to be available 24/7/365.

(6) *Customers (Stakeholders)*

Web applications can be developed for use within the boundaries of a single organisation (intranet), a number of organisations (extranets) or for use by people anywhere in the world. The implications are that stakeholders may come from a wide range of groups where some may be clearly identified (e.g., employees within an organisation) and some may remain unknown, which is often the case [4, 16, 17, 19]. As a consequence, Web developers are regularly faced with the challenge of developing applications for unknown users, whose expectations (requirements) and behaviour patterns are also unknown at development time [16]. In this case new approaches and guidelines must be devised to better understand prospective and unknown users such that quality requirements can be determined beforehand to deliver high-quality applications [19]. Whenever users are unknown it also becomes more difficult to provide aesthetically pleasing user interfaces, which are necessary to be successful and stand out from the competition [16].

Some stakeholders can reside locally, in another state/province/county, or overseas. Those who reside overseas may present different social and linguistic backgrounds, which increases the challenge of developing successful applications [4, 16]. Whenever stakeholders are unknown it is also difficult to estimate the number of users an application will service, so applications must also be scalable [17].

With regards to conventional software applications, it is usual for stakeholders be explicitly identified prior to development. These stakeholders often represent groups confined within the boundaries of departments, divisions, or organisations [16].

(7) *Update Rate (Maintenance Cycles)*

Web applications are updated frequently without specific releases and with maintenance cycles of days or even hours [17]. In addition, their content and functionality may also change significantly from one moment to another, and so the concept of project completion may seem unsuitable in such circumstances. Some organisations also allow non-information-systems experts to develop and modify Web applications, and in such environments it is often necessary to provide an overall management of the delivery and modification of applications to avoid confusion [4].

The maintenance cycle for conventional software applications complies with a more rigorous process. Upon a product's release, software organisations usually initiate a cycle whereby a list of requested changes, adjustments or improvements (either from customers or from its own development team) is prepared over a set period of time, and later incorporated as a specific version or release for distribution to all customers simultaneously. This cycle can be as short as a week and as long as several years. It requires more planning as it often entails other, possibly expensive activities such as marketing, sales, product shipping and occasionally personal installation at a customer's site [11, 17].

(8) *People involved in Development*

The Web provides a broad spectrum of different types of Web applications, varying in quality, size, complexity and technology. This variation is also applicable to the range of skills represented by those involved in Web development projects. Web applications can be created, for example, by artists and writers using simple HTML code or, more likely, one of the many commercially available Web authoring tools (e.g., Macromedia Dreamweaver, Microsoft Frontpage), making the authoring process available to those with no prior programming experience [4]. However, Web applications can also be very large and complex, requiring a team of people with diverse skills and experience. Such teams consist of Web designers and programmers, graphic designers, librarians, database designers, project managers, network security experts, and usability experts [17].

Web designers and programmers are necessary to implement the application's functionality using the necessary programming languages and technology. In particular, they also decide on the application's architecture and applicable technologies, and to design the application taking into account its documents and links [16]. Graphic designers, usability experts and librarians provide applications that are pleasing to the eye, easy to navigate and provide good search mechanisms to obtain the required information. This is often the case where such expertise is outsourced, and used on a project-by-project basis.

Large Web applications most likely use database systems for data storage, making it important to have team members with expertise in database design and the necessary queries to manipulate the data. Project managers are responsible for managing the project in a timely manner and allocating resources adequately such that applications are developed on time, within budget and are of high quality. Finally, network security experts provide solutions for various security aspects [3].

Conversely, the development of conventional software remains dominated by IT professionals, where a sound knowledge of programming, database design, and project management is necessary.

(9) *Architecture and Network*

Web applications are typically developed using a simple client–server architecture (two-tier), represented by Web browsers on client computers connecting to a Web server hosting the Web application, to more sophisticated configurations such as three-tier or even *n*-tier architectures [17]. The servers and clients within these architectures represent computers that may have different operating systems, software, hardware configurations, and may be connected to each other using different network settings and bandwidth.

The introduction of more than two tiers was motivated by limitations of the two-tier model (e.g., implementation of an application's business logic on the client machine, increased network load as any data processing is only carried out on the client machine). In such architectures the business logic is moved to a separate server (middle-tier), which services client requests for data and functionality. The middle-tier then requests and sends data to and from a (usually) separate database server. In addition, the type of networks used by the numerous stakeholders may be unknown, so assumptions have to be made while developing these Web applications [16].

Conventional software applications either run in isolation on a client machine or use a two-tier architecture whenever applications use data from database systems installed on a separate server. The type of networks used by the stakeholders is usually known in advance since most conventional software applications are limited to specific places and organisations [16].

(10) *Disciplines Involved*

A team of people with a wide range of skills and expertise in different areas is required to develop large and complex Web applications adequately. These areas reflect distinct disciplines such as software engineering (development methodologies, project management, tools), hypermedia engineering (linking, navigation), requirements engineering, usability engineering, information engineering, graphics design and network management (performance measurement and tuning) [3, 11, 19].

Building a conventional software application involves contributions from a smaller number of disciplines than those used for developing Web applications; these include software engineering, requirements engineering and usability engineering.

(11) *Legal, Social, and Ethical Issues*

The Web as a distributed environment enables a vast amount of structured (e.g., database records) and unstructured (e.g., text, images, audio) content to be easily available to a multitude of users worldwide. This is often cited as one of the greatest advantages of using the Web. However, this environment is also used for the purpose of dishonest actions, such as copying content from Web applications without acknowledging the source, distributing information about customers without their consent, infringing copyright and intellectual property rights, and even, in some instances, identity theft [16]. The consequences that follow from the unlawful use of the Web are that Web companies, customers, entities (e.g., W3C), and government agencies must apply a similar paradigm to the Web as those applied to publishing, where legal, social and ethical issues are taken into consideration [19].

Issues referring to accessibility offered by Web applications should also take into account special user groups such as the handicapped [16].

Conventional software applications also share a similar fate to that of Web applications, although to a smaller extent, since these applications are not so readily available for such a large community of users, compared to Web applications.

Table 2.1 Comparison between Web-based and traditional approaches

	Web-based approach	Traditional approach
Estimating process	Ad hoc costing of work, centred on input from the developers	More formal costing of work based on past experience from similar projects and expert opinion
Size estimation	No agreement upon a standard size measure for Web applications within the community	Lines of code or function points are the standard size measures used
Effort estimation	Effort is estimated using a bottom-up approach based on input from developers. Hardly any historical data is available from past projects	Effort is estimated using equations built taking into account project characteristics and historical data from past projects
Quality estimation	Quality is difficult to measure. Need for new quality measures specific for Web-based projects	Quality is measurable using known quality measures (e.g., defect rates, system properties)

(12) *Information Structuring and Design*

As previously mentioned, Web applications have structured and unstructured content, which may be distributed over multiple sites and use different systems (e.g., database systems, file systems, multimedia storage devices) [10]. In addition, the design of a Web application, unlike that of conventional software applications, includes the organisation of content into navigational structures by means of hyperlinks. These structures provide users with easily navigable Web applications. Well-designed applications should allow for suitable navigation structures [19], as well as the structuring of content, which should take into account its efficient and reliable management [16].

Another difference between Web and conventional applications is that Web applications often contain a variety of specific file formats for multimedia content (e.g., graphics, sound and animation). These files must be integrated into any current configuration management system, and their maintenance routines also need to be organised, as is likely that they will differ from the maintenance routines used for text-based documents [15]. Conventional software applications present structured content that uses file or database systems. The structuring of such content has been addressed by software engineering in the past so the methods employed here for information structuring and design are well known by IT professionals [16].

Reifer [18] presents a comparison between Web-based and traditional approaches that takes into account measurement challenges for project management (Table 2.1). Table 2.2 summarises the differences between Web-based and conventional development contexts.

As we have seen, there are several differences between Web development and applications and conventional development and applications. However, there are also similarities that are more evident if we focus on the development of large and complex applications. Both need quality assurance mechanisms,

Table 2.2 Web-based versus traditional approaches to development

	Web-based approach	Traditional approach
Application characteristics	Integration of numerous distinct components (e.g., fine-grained, interpreted scripting languages, COTS, multimedia files, HTML/SGML/XML files, databases, graphical images), distributed, cross-platform applications and structuring of content using navigational structures with hyperlinks	Integration of distinct components (e.g., COTS, databases, graphical images), monolithic single-platform applications
Primary technologies used	Variety of Java solutions (Java servlets, Enterprise JavaBeans, applets, and JavaServer Pages), HTML, JavaScript, XML, UML, databases, third-party components and middleware, etc.	Object-oriented methods, generators, and languages, relational databases, and CASE tools
Approach to quality delivered	Quality is a higher priority than time to market	Time to market takes priority over quality
Development process drivers	Reliability, usability and security	Time to market
Availability of the application	Throughout the whole year (24/7/365)	Except for a few application domains, no need for availability 24/7/365
Customers (stakeholders)	Wide range of groups, known and unknown, residing locally or overseas	Generally groups confined within the boundaries of departments, divisions, or organizations
Update rate (maintenance cycles)	Frequently without specific releases, maintenance cycles of days or even hours	Specific releases, maintenance cycles ranging from a week to several years
People involved in development	Web designers and programmers, graphic designers, librarians, database designers, project managers, network security experts, usability experts, artists, writers	IT professionals with knowledge of programming, database design and project management
Architecture and Network	Two-tier to n-tier clients and servers with different network settings and bandwidth, sometimes unknown	One- to two-tier architecture, network settings and bandwidth are likely to be known in advance
Disciplines involved	Software engineering, hypermedia engineering, requirements engineering, usability engineering, information engineering, graphics design and network management	Software engineering, requirements engineering, and usability engineering
Legal, social, and ethical issues	Content can be easily copied and distributed without permission or acknowledgement of copyright and intellectual property rights. Applications should take into account all groups of users including those handicapped	Content can also be copied infringing privacy, copyright, and IP issues, albeit to a smaller extent
Information structuring and design	Structured and unstructured content, use of hyperlinks to build navigational structures	Structured content, infrequent use of hyperlinks

development methodologies, tools, processes, techniques for requirements elicitation, effective testing and maintenance methods, and tools [19].

Conclusions

This chapter discussed differences between Web and software applications, and their development processes based on the following 12 areas:

1. Application characteristics
2. Primary technologies used
3. Approach to quality delivered
4. Development process drivers
5. Availability of the application
6. Customers (stakeholders)
7. Update rate (maintenance cycles)
8. People involved in development
9. Architecture and network
10. Disciplines involved
11. Legal, social, and ethical issues
12. Information structuring and design

Acknowledgements We would like to thank Tayana Conte for her comments on a previous version of this chapter.

References

1. Murugesan S, Deshpande Y (2002) Meeting the challenges of web application development: the web engineering approach. In: Proceedings of the 24th international conference on software engineering, Orlando, FL, pp 687–688, May 2002
2. Gellersen H, Wicke R, Gaedke M (1997) WebComposition: an object-oriented support system for the Web engineering lifecycle. J Comput Netw ISDN Syst 29(8–13):865–1553 [also (1996) In: Proceedings of the sixth international world wide web conference, pp 429–1437]
3. Ginige A (2002) Workshop on web engineering: Web engineering: managing the complexity of Web systems development. In: Proceedings of the 14th international conference on software engineering and knowledge engineering, New York, NY, pp 721–729, Jul 2002
4. Standing C (2002) Methodologies for developing Web applications. Inf Softw Technol 44(3): 151–160
5. Collins English Dictionary (2000) HarperCollins
6. Azhar D, Mendes E, Riddle P (2012) A systematic review of web resource estimation. In: Proceedings of promise'12, New York, pp 49–58
7. Taylor MJ, McWilliam J, Forsyth H, Wade S (2002) Methodologies and website development: a survey of practice. Inf Softw Technol 44(6):381–391
8. Ricca F, Tonella P (2001) Analysis and testing of Web applications. In: Proceedings of the 23rd international conference on software engineering, pp 25–34
9. Lee SC, Shirani AI (2004) A component based methodology for Web application development. J Syst Softw 71(1–2):177–187
10. Fraternali P, Paolini P (2000) Model-driven development of Web applications: the AutoWeb system. ACM Trans Inform Syst 18(4):1–35
11. Ginige A, Murugesan S (2001) Web engineering: an introduction. IEEE Multimed 8(1):14–18

12. Murugesan S, Deshpande Y (2001) Web engineering, managing diversity and complexity of web application development, Lecture notes in computer science 2016. Springer, Heidelberg
13. IEEE Std. 2001-2002 (2003) Recommended practice for the internet web site engineering, web site management, and web site life cycle. IEEE
14. Christodoulou SP, Zafiris PA, Papatheodorou TS (2000) WWW 2000: the developer's view and a practitioner's approach to web engineering. In: Proceedings of the 2nd ICSE workshop on web engineering, Limerick, pp 75–92
15. Brereton P, Budgen D, Hamilton G (1998) Hypertext: the next maintenance mountain. Computer 31(12):49–55
16. Deshpande Y, Hansen S (2001) Web engineering: creating a discipline among disciplines. IEEE Multimed 8(2):8–87
17. Offutt J (2002) Quality attributes of Web software applications. IEEE Softw 19(2):25–32
18. Reifer DJ (2000) Web development: estimating quick-to-market software. IEEE Softw 17(6): 57–64
19. Deshpande Y, Murugesan S, Ginige A, Hansen S, Schwabe D, Gaedke M, White B (2002) Web engineering. J Web Eng 1(1):3–17

Introduction to Effort Estimation

3

Introduction

Web development is a relatively new but rapidly growing industry where numerous companies use the Web as a delivery platform for a diverse range of applications, from complex e-commerce solutions with back-end databases to on-line personal static Web pages and blogs. With the sheer diversity relating to both the types of Web applications and the technologies employed to develop them, there is an ever-growing number of companies bidding for as many Web projects as they can accommodate. As usual, in order to win the bids, companies opt to estimate unrealistic schedules, leading to applications that are rarely developed on time and within budget.

Many reasons can lead to unrealistic estimates and schedules; these include the lack of understanding about the basic building blocks that are part of an effort estimation process. Therefore, in this chapter we introduce and detail these building blocks, using Web development and project characteristics as examples.

It should be noted that cost and effort are often used interchangeably within the context of existing literature in effort estimation as effort is often taken by project managers as the main component of project costs. However, given that project costs also take into account other factors such as contingency and profit [1], we use the word "effort" and not "cost" throughout this chapter and book, in order to solely represent the amount of effort in person-hours that is needed in order to develop a Web application.

An Overview of the Effort Estimation Process

The purpose of estimating effort as part of managing a project is to predict the amount of effort required to accomplish the set of tasks needed as part of a project's life cycle, based on a set of inputs such as the knowledge/data of previous "similar"

E. Mendes, *Practitioner's Knowledge Representation*, DOI 10.1007/978-3-642-54157-5_3,   27
© Springer-Verlag Berlin Heidelberg 2014

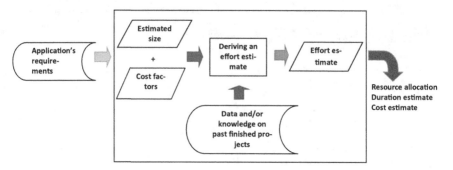

Fig. 3.1 Effort estimation process

projects and other application and project characteristics that are believed to be related to effort.

Using a black box metaphor, application and project characteristics (independent variables) and data/knowledge from past projects are the input into an effort estimation process, and an effort estimate (dependent variable) is the output we wish to predict. Figure 3.1 shows an extended view of an effort estimation process comprising not only direct inputs into and output from the process, but also some other related input and outputs. Each of these parts is detailed next:

1. Application's requirements: represents any set of requirements—functional and non-functional—that were obtained from elicitation meetings with clients. They may be detailed using natural language or some other notation such as UML use cases.
2. Estimated size: represents an estimate relating to the size of the "problem" to be solved, which once implemented will represent a delivered application. Possible examples are *an estimated number of new Web pages* and *a number of functions/features (e.g., shopping cart) to be offered by the new Web application*.
3. Cost factors: represent factors that are not size-related but that are believed to be associated with effort in the sense that they have an effect upon the total amount of effort necessary to develop a Web application. Possible examples are *the total number of team members who will participate in the development of the new Web application, developers' average number of years of experience with the development tools that will be employed to implement the Web application, the project manager's previous experience managing a similar project* and *the nature of the client who requested the Web application (e.g., grumpy, not grumpy)*.
4. Data and/or knowledge on past finished projects: represents either or both of (a) hard data on past finished projects gathered by the company, (b) expert knowledge from project managers and developers for previous projects that are somewhat similar to the one for which effort needs to be estimated.
5. Effort estimate: the total estimated effort (generally measured in person-hours) that is needed to complete the project and deliver the Web application.

6. Resource allocation: represents the process of deciding upon the resources (e.g., developers, testers, tools) that need to be allocated to the project as a result of the effort that was estimated. This also needs to take into account other existing projects being currently managed (project portfolio).
7. Duration estimate: represents an estimate of the duration (calendar periods) by when the project will be completed. This also needs to take into account other existing projects being currently managed (project portfolio).
8. Cost estimate: estimated cost of the project that is based on the estimated effort, plus contingency and profit costs.

The process of deriving an effort estimate can be described as follows:

1. Data relating to the new application for which effort is to be estimated (estimated size and cost factors) is used as input to the process.
2. Data and/or knowledge on past finished projects, for which actual effort is known, are used by project managers and developers in order to identify any similarities between applications developed previously and the new application described in (1) above. Whenever a company does not have either data or experience on similar projects and corresponding applications, they use instead an "educated guess" based on prior experience with dissimilar projects.
3. The output of this process is an effort estimate (dependent variable), which is then used to allocate resources and to estimate project duration and costs.

Regardless of the type of application for which effort is to be estimated, a general rule is that the one consistent input (independent variable) believed to have the strongest influence on effort is size (i.e., the total number of Web pages), with cost drivers coming second but also playing an influential role. Finally, each of the tasks to be estimated can be as simple as developing a single function (e.g., creating a Web form with ten fields) or as complex as developing a script to communicate with a back-end relational database system.

Several techniques for effort estimation have been proposed over the past 30 years in software engineering, and over the past 12 years in Web engineering. These fall into three broad categories [2]: expert-based effort estimation, algorithmic models and artificial intelligence techniques. Each category is described in the next sections.

Expert-Based Effort Estimation

Expert-based effort estimation is the process of estimating effort by subjective means, and is often based on previous experience with developing and/or managing similar projects. This is by far the most commonly used technique for Web effort estimation, with the attainment of accurate effort estimates being directly proportional to the competence and experience of the individuals involved (e.g., project manager, developer). Within the context of Web development, our experience suggests that expert-based effort estimates are obtained using one of the following mechanisms:

- An estimate that is based on a detailed effort breakdown that takes into account all of the lowest-level parts of an application and the functional tasks necessary to develop this application. Each task attributed with effort estimates is repeatedly combined into higher-level estimates until we finally obtain one estimate that is considered as the sum of all lower-level estimate parts. This type of estimation is called bottom-up. Each estimate can be an educated guess or can be based on sound previous experience with similar projects.
- An estimate representing an overall process to be used to produce an application, as well as knowledge about the application to be developed, i.e., the product. A total estimate is suggested and used to calculate estimates for the component parts (tasks), relative portions of the whole. This type of estimation is called top-down.

Estimates can be suggested by a project manager, or by a group of people mixing project manager(s) and developers, usually by means of a brainstorming session.

A survey of 32 Web companies in New Zealand conducted in 2004 [3] showed that 32 % prepared effort estimates during the requirements gathering phase, 62 % prepared an estimates during their design phase, while 6 % did not have to provide any effort estimates to their customers since they were happy to pay for the development costs without the need for a quote.

Of the 32 companies surveyed, 38 % did not refine their effort estimate, and 62 % did refine their estimates but not often. For the companies surveyed, this indicates that for the majority of companies, the initial effort estimate was used as their "final" estimate, and work was adjusted to fit this initial quote. These results corroborated those published in [4].

Sometimes Web companies gather data on effort for past Web projects believing this data is sufficient to help obtain accurate estimates for new projects. However, without understanding the factors that influence effort within the context of a specific company, effort data alone is unlikely to be sufficient to warrant successful results.

The drawbacks of expert-based estimation can be identified as follows:

1. It is very difficult to quantify and to clearly determine the factors that have been used to derive an estimate, making it difficult to apply the same reasoning to other projects (repeatability);
2. When a company finally builds up its expertise with developing Web applications, using a given set of technologies, other technologies appear and are rapidly adopted (mostly due to hype), thus leaving behind valuable knowledge from the past.
3. Obtaining an effort estimate based on experience with past similar projects can be misleading when projects vary in their characteristics. For example, knowing that a Web application containing ten new static HTML, ten new images with a development time of 40 person-hours does not mean that a similar application developed by two people will also consume 40 person-hours to complete the

task. Two people may need additional time to communicate, and may also have different experience with using HTML. In addition, another application eight times its size is unlikely to take exactly eight times longer to complete. This suggests that experience alone is not enough to identify the underlying relationship between effort and size/cost drivers (e.g., linear or exponential).

4. Developers and project managers are known for providing optimistic effort estimates [5] for tasks that they have to carry out themselves. Optimistic estimates lead to underestimated effort with the direct consequence of projects being over budget and over time.

To cope with underestimation, it is suggested that experts provide three different estimates [6]: an optimistic estimate o, a realistic estimate r, and a pessimistic estimate p. Based on a beta distribution, the estimated effort E is then calculated as:

$$E = (o + 4r + p)/6 \tag{3.1}$$

This measure is likely to be better than a simple average of o and p; however, caution is still necessary.

Although there are problems related to using expert-based estimations, few studies have reported that when used in combination with other less subjective techniques (e.g., algorithmic models) expert-based effort estimation can be an effective estimating tool [7, 8].

Expert-based effort estimation is a process that has not been objectively detailed; however, it can still be represented in terms of the diagram presented in Fig. 3.1, where the order of steps that take place to obtain an expert-based effort estimate are as follows:

Step 1. An expert/group of developers implicitly look(s) at the estimated size and cost drivers related to a new project for which effort needs to be estimated.

Step 2. Based on the data obtained in Step 1, they/(s)he remember(s) or retrieve(s) data/knowledge on past finished projects for which actual effort is known.

Step 3. Based on the data from Steps 1 and 2, they/(s)he subjectively estimate(s) effort for the new project.

In summary, when employing an expert-based approach to effort estimation, the knowledge regarding the characteristics of a new project is used to retrieve, from either memory or a database, knowledge on finished similar projects. Once this knowledge is retrieved, effort can be estimated.

It is important to stress that within a context where estimates are obtained via expert-based opinion, deriving a good effort estimate is much more likely to occur when the previous knowledge/data about completed projects relates to projects that are very similar to the one having its effort estimated. Here we use the principle "similar problems have similar solutions". Note that for this assumption to be correct we also need to guarantee that the productivity of the team working on the new project is similar to the productivity of the team(s) for the past similar projects.

The problems aforementioned related to expert-based effort estimation led to the proposal of techniques aimed to formalise the effort estimation process. Such techniques are presented in the next sections.

Algorithmic Techniques

Algorithmic techniques are the most popular techniques described in the Web and software effort estimation literature. Such techniques attempt to build models that precisely represent the relationship between effort and one or more project characteristics via the use of algorithmic models. Such models assume that application size is the main contributor to effort; thus in any algorithmic model the central project characteristic used is usually taken to be some notion of application size (e.g., the number of lines of source code, function points, number of Web pages, number of new images). The relationship between size and effort is often translated into an equation shown by Eq. (3.2), where a and b are constants, S is the estimated size of an application, and E is the estimated effort required to develop an application of size S.

$$E = a \ S^b \qquad (3.2)$$

In Eq. (3.2), when $b < 1$ we have economies of scale, i.e., larger projects use less effort, comparatively, than smaller projects. The opposite situation ($b > 1$) gives diseconomies of scale, i.e., larger projects use more effort, comparatively, than smaller projects. When b is either $>$ or < 1, the relationship between S and E is nonlinear. Conversely, when $b = 1$ the relationship is linear.

However, size alone is unlikely to be the only contributor to effort. Other project characteristics, such as developer's programming experience, tools used to implement an application and maximum or average team size are also believed to influence the amount of effort required to develop an application. As previously said, these variables are known in the literature as *cost drivers*. Therefore, an algorithmic model should include not only size but also the cost drivers believed to influence effort. Thus effort is determined mainly by size; however, its value can also be adjusted by taking into account cost drivers (Eq. 3.3).

$$E = a \ S^b CostDrivers \qquad (3.3)$$

Different proposals have been made in an attempt to define the exact form such algorithmic models should take. The most popular are presented next.

COCOMO

One of the first algorithmic models to be proposed in the literature was the Constructive COst MOdel (COCOMO) [9]. COCOMO aimed to be a generic algorithmic model that could be applied by any organisation to estimate effort at three different stages in the development life cycle of a software project: early on in the development life cycle, when requirements have not yet been fully specified (Basic COCOMO); once detailed requirements have been specified (Intermediate COCOMO); and when the application's design has been finalised (Advanced COCOMO). Each stage corresponds to a different model, and all three models take the same form (Eq. 3.3):

$$EstimatedEffort = a \ EstSizeNewproj^b \ EAF \qquad (3.4)$$

where:

- *EstimatedEffort* is the estimated effort, measured in person-months, to develop an application;
- *EstSizeNewproj* is the size of an application measured in thousands of delivered source instructions (KDSI);
- *a* and *b* are constants which are determined by the class of project to be developed. The three possible classes are:
  - *Organic*: The *organic* class incorporates small, noncomplicated software projects, developed by teams that have a great deal of experience with similar projects, and where software requirements are not strict.
  - *Semidetached*: The *semidetached* class incorporates software projects that are halfway between small-to-easy and large-to-complex. Development teams show a mix of experiences, and requirements also present a mix of strict and slightly vague requirements.
  - *Embedded*: The *embedded* class incorporates projects that must be developed within a context where there are rigid hardware, software and operational restrictions.
- *EAF* is an effort adjustment factor, calculated from cost drivers (e.g., developers, experience, tools).

The COCOMO model makes it clear that size is the main component of an effort estimate. Constants *a* and *b*, and the adjustment factor *EAF* all vary depending on the model used, and in the following ways:

The Basic COCOMO uses an value *EAF* of 1; *a* and *b* differ depending on a project's class (Table 3.1).

The Intermediate COCOMO calculates *EAF* based on 15 cost drivers, grouped into four categories: product, computer, personnel and project (Table 3.2). Each cost driver is rated on a 6-point ordinal scale ranging from "very low importance" to "extra high importance". Each scale rating determines an effort multiplier, and the product of all 15 effort multipliers is taken as the *EAF*.

Table 3.1 Parameter
values for basic and
intermediate COCOMO

	Class	a	b
Basic	Organic	2.4	1.05
	Semidetached	3.0	1.12
	Embedded	3.6	1.20
Intermediate	Organic	3.2	1.05
	Semidetached	3.0	1.12
	Embedded	2.8	1.20

Table 3.2 Cost drivers
used in the intermediate
and advanced COCOMO

	Cost driver
Personnel	Analyst capability
	Applications experience
	Programmer capability
	Virtual machine experience
	Language experience
Project	Modern programming practices
	Software tools
	Development schedule
Product	Required software reliability
	Database size
	Product complexity
Computer	Execution time constraint
	Main storage constraint
	Virtual machine volatility
	Computer turnaround time

Table 3.3 Example of rating in the advanced COCOMO

Cost driver	Rating	RPD	DD	CUT	IT
ACAP (analyst CAPability)	Very low	1.8	1.35	1.35	1.5
	Low	0.85	0.85	0.85	1.2
	Nominal	1	1	1	1
	High	0.75	0.9	0.9	0.85
	Very high	0.55	0.75	0.75	0.7

The Advanced COCOMO uses the same 15 cost drivers as the Intermediate
COCOMO; however, they are all weighted according to each phase of the development lifecycle, i.e., each cost driver is broken down by development phase (see
example in Table 3.3). This model therefore enables the same cost driver to be rated
differently depending on the development phase. In addition, it views a software
application as a composition of modules and subsystems to which the Intermediate
COCOMO model is applied.

The four development phases used in the Advanced COCOMO model are
requirements planning and product design (RPD), detailed design (DD), coding
and unit test (CUT), and integration and test (IT). An overall project estimate is

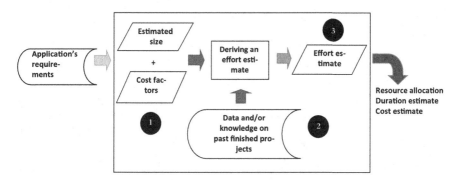

Fig. 3.2 Expert-based effort estimation

obtained by aggregating estimates obtained for subsystems, which themselves were obtained by combining estimates made for each module.

The original COCOMO model was radically improved 15 years later and renamed as COCOMO II model, which incorporates knowledge of changes that have occurred in software development environments and practices over the previous 15 years [10]. COCOMO II is not detailed in this book; however, interested readers are referred to [10, 11].

The COCOMO model is an example of a *general purpose* model, where it is assumed that it is not compulsory for ratings and parameters to be adjusted (calibrated) to specific companies in order for the model to be used effectively.

Despite the existence of *general purpose* models, such as COCOMO, the effort estimation literature has numerous examples of *specialised* algorithmic models that were built using applied regression analysis techniques [12] on data sets of past completed projects. Specialised and regression-based algorithmic models are most suitable to local circumstances, such as "in-house" analysis, as they are derived from past data that often represents projects from the company itself. Regression analysis, used to generate regression-based algorithmic models, provides a procedure for determining the "best" straight-line fit (Fig. 3.2) to a set of project data that represents the relationship between effort (response or dependent variable) and cost drivers (predictor or independent variables) [12].

Figure 3.2 shows, using real data on Web projects, an example of a regression line that describes the relationship between log(*Effort*) and log(*totalWebPages*). It should be noted that the original variables *Effort* and *totalWebPages* have been transformed using the natural logarithmic scale to comply more closely with the assumptions of the regression analysis techniques.

The equation represented by the regression line in Fig. 3.3 is as follows:

$$\log Effort = \log a + b \log totalWebPages \qquad (3.5)$$

where log a is the point in which the regression line intercepts the Y-axis, known simply as the *intercept*, and b represents the slope of the regression line, i.e., its inclination, generically represented by the form,

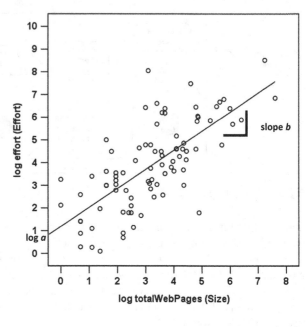

Fig. 3.3 Example of a regression line

$$y = mx + c \qquad (3.6)$$

Equation (3.5) shows a linear relationship between log(*Effort*) and log (*totalWebPages*). However, since the original variables have been transformed before the regression technique was employed, this equation needs to be transformed back such that it uses the original variables. The resultant equation is:

$$Effort = a \ totalWebPages^b \qquad (3.7)$$

Other examples of equations representing regression lines are given in Eqs. (3.8) and (3.9):

$$EstimatedEffort = C + a_0 EstSizeNewproj + a_1 CD_1 + \cdots + a_n CD_n \qquad (3.8)$$

$$EstimatedEffort = C \ EstSizeNewproj^{a_0} \ CD_1{}^{a_1} \cdots CD_n{}^{a_n} \qquad (3.9)$$

where C is the regression line's intercept, a constant denoting the initial estimated effort (assuming size and cost drivers to be zero), $a_0 \ldots a_n$ are constants derived from past data, and $CD_1 \ldots CD_n$ are cost drivers that have an impact on effort.

Regarding the regression analysis itself, two of the most widely used techniques are multiple regression (MR) and stepwise regression (SWR). The difference between these two techniques is that MR obtains a regression line using all the independent variables at the same time, whereas SWR is a technique that examines different combinations of independent variables, looking for the best grouping to explain the greatest amount of variation in effort. Both use least-squares regression, where the regression line selected is the one that reflects the minimum values of the

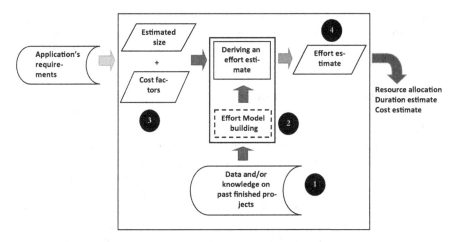

Fig. 3.4 Using an algorithmic technique for effort estimation

sum of the squared errors. Errors are calculated as the difference between actual and estimated effort and are known as "residuals" [12].

In terms of the diagram presented in Fig. 3.4, an algorithmic model uses constant scalar values based on past project data; however, for anyone wishing to use this model, the steps to use are 1, 2, 3 and 4.

A *general purpose* algorithmic model tends to use Step 2 once, and use the values obtained for all its constants, estimated size and cost drivers to derive effort estimates. It is common for such models to be used by companies without recalibration of values for the constants. Within the context of a *specialised* algorithmic model, Step 2 is used whenever it is necessary to recalibrate the model. This can occur after several new projects are finished and incorporated to the company's database of data on past finished projects. However, a company may also decide to recalibrate a model after every new project is finished, or to use the initial model for a longer time period. If the development team remains unchanged (and assuming that the team does not have an excessive learning curve for each new project) and new projects are similar to past projects, there is no pressing need to recalibrate an algorithmic model too often.

The sequence of steps (Fig. 3.4) is as follows:

Step 1. Past data is used to generate an algorithmic model.
Step 2. An algorithmic model is built from past data obtained in Step 1.
Step 3. The model created in Step 2 then receives, as input, values for the estimated size and cost drivers relative to the new project for which effort is to be estimated.
Step 4. The model generates an estimated effort.

The above-mentioned sequence differs from that for expert opinion, shown in Fig. 3.2.

In summary, when employing an algorithmic technique to estimating effort, an effort estimate for a new project can only be obtained after building an algorithmic model from past data, and using it with the estimated size and cost drivers.

Artificial Intelligence Techniques

Artificial intelligence techniques have, in the past 20 years been used as a complement to, or as an alternative to, the previous two categories. Examples include fuzzy logic [13], regression trees [14], neural networks [15] and case-based reasoning [2]. We will cover case-based reasoning (CBR) and regression trees (CART) in more detail as these have been to date the most popular machine learning techniques employed for Web effort estimation. A useful summary of numerous machine learning techniques can also be found in [16]. Note that there is yet another machine learning technique that will not be covered in this chapter, as it will be detailed separately in Chap. 5, given this is the technique focus of this book.

Case-based reasoning (CBR): CBR uses the assumption that *similar problems provide similar solutions*. It provides estimates by comparing the characteristics of the current project to be estimated, against a library of historical information from completed projects with known effort (case base).

Using CBR involves [17]:

1. Characterising a new project p, for which an estimate is required, with variables (features) common to those completed projects stored in the case base. In terms of Web and software effort estimation, features represent size measures and cost drivers that have a bearing on effort. This means that, if a Web company has stored data on past projects where, for example, the data represents the features *effort*, *size*, *development team size* and *tools used*, the data used as input to obtaining an effort estimate will also need to include these same features.
2. Use of this characterisation as a basis for finding similar (analogous) completed projects, for which effort is known. This process can be achieved by measuring the "distance" between two projects at a time (project p and one finished project), based on the features' values, for all features (k) characterising these projects. Each finished project is compared to project p, and the finished project presenting the shortest distance overall is the "most similar project" to project p. Although numerous techniques can be used to measure similarity, nearest neighbour algorithms using the unweighted Euclidean distance measure have been the most widely used to date in Web and software engineering.
3. Generation of a predicted value of effort for project p based on the effort for those completed projects that are similar to p. The number of similar projects taken into account to obtain an effort estimate will depend on the size of the case base. For small case bases (e.g., up to 90 cases), typical values use the most similar finished project, or the two or three most similar finished projects (1, 2 and 3 closest neighbours/analogues). For larger case bases no conclusions have been reached regarding the best number of similar projects to use. The calculation of estimated effort is obtained using the same effort value as the closest

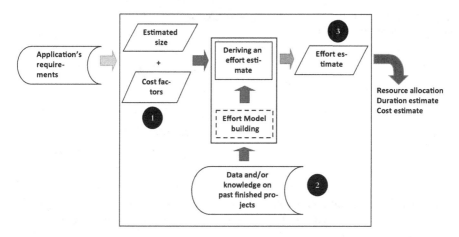

Fig. 3.5 Using CBR for effort estimation

neighbour, or the mean effort for two or more closest neighbours. This is the common choice in Web and software engineering.

Figure 3.5 shows the sequence of steps used with CBR in order to obtain an effort estimate:

Step 1. The estimated size and cost drivers relating to a new project p are used as input to retrieve similar projects from the case base, for which actual effort is known.

Step 2. Using the data from Step 1, a suitable CBR tool retrieves similar projects to project p, and ranks these similar projects in ascending order of similarity, i.e., from "most similar" to "least similar".

Step 3. A suitable CBR tool calculates estimated effort for project p.

The sequence just described is similar to the one employed when obtaining estimated effort using expert opinion. Both require that the characteristics of a new project be known in order to retrieve similar finished projects. Once similar projects are retrieved, effort can be estimated.

When using CBR there are six parameters that need to be considered, which are as follows [18]:

Feature Subset Selection

Feature subset selection involves determining the optimum subset of features that yields the most accurate estimation. Some existing CBR tools, e.g., ANGEL [2], optionally offer this functionality using a brute force algorithm, searching for all possible feature subsets. Other CBR tools (e.g., CBR-Works from tec:inno) have no

Fig. 3.6 Euclidean distance
using two size features ($n = 2$)

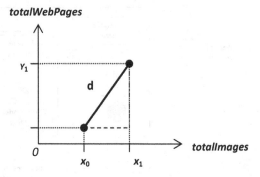

such functionality, and therefore to obtain estimated effort, we must use all of the
known features of a new project to retrieve the most similar finished projects.

Similarity Measure

The similarity measure records the level of similarity between different cases.
Several similarity measures have been proposed in the literature to date; the three
most popular currently used in the Web and software engineering literature [17–19]
are the unweighted Euclidean distance, the weighted Euclidean distance and the
maximum distance. However, there are also other similarity measures available,
which are presented in [17]. Each of the three similarity afore-mentioned measures
is described below.

Unweighted Euclidean Distance The unweighted Euclidean distance measures
the Euclidean (straight-line) distance d between two cases, where each case has
n features. The equation used to calculate the distance between two cases x and y is
the following:

$$d(x, y) = \sqrt{|x_0 - y_0|^2 + |x_1 - y_1|^2 + \ldots + |x_{n-1} - y_{n-1}|^2 + |x_n - y_n|^2} \quad (3.10)$$

where x_0 to x_n represent features 0 to n of case x; y_0 to y_n represent features 0 to n of
case y.

This measure has a geometrical meaning as the shortest distance between two
points in an n-dimensional Euclidean space [17] (Fig. 3.6).

Figure 3.6 illustrates the unweighted Euclidean distance by representing
coordinates in a two-dimensional space, E2 as the number of features employed
determines the number of dimensions, En.

Given the following example:

Project	totalWebPages	totalImages
1 (new)	100	20
2	350	12
3	220	25

The unweighted Euclidean distance between the new project 1 and finished project 2 would be calculated using the following equation:

$$d = \sqrt{|100 - 350|^2 + |20 - 12|^2} = 250.128 \qquad (3.11)$$

The unweighted Euclidean distance between the new project 1 and finished project 3 would be calculated using the following equation:

$$d = \sqrt{|100 - 220|^2 + |20 - 25|^2} = 120.104 \qquad (3.12)$$

Using the weighted Euclidean distance, the distance between projects 1 and 3 is smaller than the distance between projects 1 and 2; thus project 3 is more similar than project 2 to project 1.

Weighted Euclidean Distance The weighted Euclidean distance is used when features are given weights that reflect the relative importance of each feature. The weighted Euclidean distance measures the Euclidean distance d between two cases, where each case has n features and each feature has a weight w. The equation used to calculate the distance between two cases x and y is the following:

$$d(x, y) = \sqrt{w_0|x_0 - y_0|^2 + w_1|x_1 - y_1|^2 + \ldots + w_{n-1}|x_{n-1} - y_{n-1}|^2 + w_n|x_n - y_n|^2}$$
$$(3.13)$$

where x_0 to x_n represent features 0 to n of case x; y_0 to y_n represent features 0 to n of case y; w_0 to w_n are the weights for features 0 to n.

Maximum Distance The maximum distance computes the highest feature similarity, i.e., the one to define the closest analogy. For two points (x_0, y_0) and (x_1, y_1), the maximum measure d is equivalent to the equation:

$$d = \sqrt{\max\left((x_0 - y_0)^2, (x_1 - y_1)^2\right)} \qquad (3.14)$$

This effectively reduces the similarity measure down to a single feature, although this feature may differ for each retrieval episode. So, for a given "new" project P_{new}, the closest project in the case will be the one that has at least one size feature with the most similar value to the same feature in project P_{new}.

Scaling

Scaling (also known as standardisation) represents the transformation of a feature's values according to a defined rule, such that all features present values within the same range and as a consequence have the same degree of influence on the result [17]. A common method of scaling is to assign 0 to the observed minimum value and 1 to the maximum observed value [20], a strategy used by ANGEL and CBR-Works. Original feature values are normalised (between 0 and 1) by case-based reasoning tools to guarantee that they all influence the results in a similar fashion.

Number of Analogies

The number of analogies refers to the number of most similar cases that will be used to generate an effort estimate. With small sets of data, it is reasonable to consider only a small number of the most similar analogues [17]. Several studies in Web and software engineering have used only the closest case/analogue ($k = 1$) to obtain an estimated effort for a new project [21, 22], while others have also used the two closest and the three closest analogues [17, 19, 23–28].

Analogy Adaptation

Once the most similar cases have been selected the next step is to identify how to generate (adapt) an effort estimate for project P_{new}. Choices of analogy adaptation techniques presented in the literature vary from the nearest neighbour [21, 24], the mean of the closest analogues [2, 29], the median of the closest analogues [17], the inverse distance-weighted mean and inverse rank weighted-mean [20], to illustrate just a few. The adaptations used to date for Web engineering are the nearest neighbour, mean of the closest analogues [19, 25] and the inverse rank-weighted mean [22, 26, 27, 30].

Each adaptation is explained below:

Nearest Neighbour For the estimated effort P_{new}, this type of adaptation uses the same effort of its closest analogue.

Mean Effort For the estimated effort P_{new}, this type of adaptation uses the average of its closest k analogues, when $k > 1$. This is a typical measure of central tendency, often used in the Web and software engineering literature. It treats all analogues as being equally important towards the outcome—the estimated effort.

Median Effort For the estimated effort P_{new}, this type of adaptation uses the median of the closest k analogues, when $k > 2$. This is also a measure of central tendency, and has been used in the literature when the number of selected closest projects is >2 [17].

Inverse Rank Weighted Mean This type of adaptation allows higher ranked analogues to have more influence over the outcome than lower ones. For example, if we use three analogues, then the closest analogue (CA) would have weight $= 3$, the second-closest analogue (SC) would have weight $= 2$ and the third closest analogue (LA) would have weight $= 1$. The estimated effort would then be calculated as:

$$Inverse\,Rank\,Weighed\,Mean = \frac{3CA + 2SC + LA}{6}$$ (3.15)

Adaptation Rules

Adaptation rules are used to adapt the estimated effort, according to a given criterion, such that it reflects the characteristics of the target project (new project) more closely. For example, in the context of effort prediction, the estimated effort to develop an application a would be adapted such that it would also take into consideration the size value of application a. The adaptation rule that has been employed to date in Web engineering is based on the linear size adjustment to the estimated effort [26, 27], obtained as follows:

- Once the most similar analogue in the case base has been retrieved, its effort value is adjusted and used as the effort estimate for the target project (new project).
- A linear extrapolation is performed along the dimension of a single measure, which is a size measure strongly correlated with effort. The linear size adjustment is calculated using the equation presented below.

$$Effort_{newProject} = \frac{Effort_{finishedProject}}{Size_{finishedProject}} Size_{newProject} \qquad (3.16)$$

Given the following example:

Project	totalWebPages (size)	totalEffort (effort)
Target (new)	100 (estimated value)	20 (estimated and adapted value)
Closest analogue	350 (actual value)	70 (actual value)

The estimated effort for the target project will be calculated as:

$$Effort_{newProject} = \frac{70}{350} 100 = 20 \qquad (3.17)$$

When we use more than one size measure as feature, the equation changes to:

$$E_{est.P} = \frac{1}{q} \left(\sum_{q=1}^{q=x} \frac{E_{act} S_{est.q}}{S_{act.q}\big|_{>0}} \right) \qquad (3.18)$$

where:
  q is the number of size measures used as features.
  $E_{est.P}$ is the total effort estimated for the new Web project P.
  E_{act} is the total effort for the closest analogue obtained from the case base.
  $S_{est.q}$ is the estimated value for the size measure q, which is obtained from the client.
  $S_{act.q}$ is the actual value for the size measure q, for the closest analogue obtained from the case base.
  This type of adaptation assumes that all projects present similar productivity; however, it may not necessarily represent the Web development context of numerous Web companies worldwide.

Classification and Regression Trees

Classification and Regression Trees (CART) [31] use independent variables (predictors) to build binary trees, where each leaf node represents either a category to which an estimate belongs, or a value for an estimate. The former situation occurs with *classification trees,* and the latter occurs with *regression trees,* i.e., whenever predictors are categorical (e.g., Yes/No) the CART tree is called a *classification tree,* and whenever predictors are numerical the CART tree is called a *regression tree.*

In order to obtain an estimate one has to traverse tree nodes from root to leaf by selecting the nodes that represent the category or value for the independent variables associated with the project to be estimated.

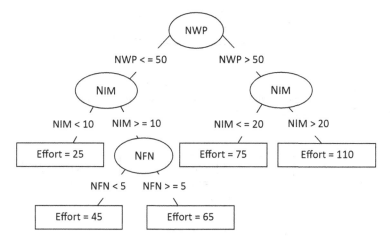

Fig. 3.7 Example of a regression tree for Web effort estimation

For example, assume we wish to obtain an effort estimate for a new Web project using as its basis the simple regression tree structure presented in Fig. 3.7. This regression tree was generated from data obtained from past completed Web applications, taking into account their existing values of effort and independent variables (e.g., new Web pages (NWP), new images (NIM), and new features/functions (NFN)). The data used to build a CART model is called a *learning sample*, and once a tree has been built it can be used to estimate effort for new projects. Assuming that the estimated values for NWP, NIM and NFN for a new Web project are 25, 15 and 3, respectively, we would obtain an estimated effort of 45 person-hours after navigating the tree from its root down to leaf "Effort = 45".

If we now assume that the estimated values for NWP, NIM and NFN for a new Web project are 56, 34 and 22, respectively, we would obtain an estimated effort of 110 person-hours after navigating the tree from its root down to leaf "Effort = 110".

A simple example of a classification tree for Web effort estimation is depicted in Fig. 3.8. It uses the same variable names as those shown in Fig. 3.7, however, these variables are now all categorical, where possible categories (classes) are "Yes" and "No". The effort estimate obtained using this classification tree is also categorical, where possible categories are "High effort" and "Low effort".

A CART model constructs a binary tree by recursively partitioning the predictor space (set of all values or categories for the independent variables judged relevant) into subsets where the distribution of values or categories for the dependent variable (e.g., effort) is successively more uniform. The partition (split) of a subset *S1* is decided on the basis that the data in each of the descendant subsets should be "purer" than the data in *S1*. Thus node "impurity" is directly related to the amount of different values or classes in a node, i.e., the greater the mix of classes or values, the higher the node "impurity". A "pure" node means that all the cases (e.g., Web projects) belong to the same class, or have the same value. The partition of subsets continues until a node contains only one class or value. Note that it is not

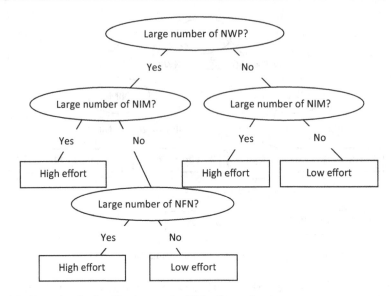

Fig. 3.8 Example of a classification tree for Web effort estimation

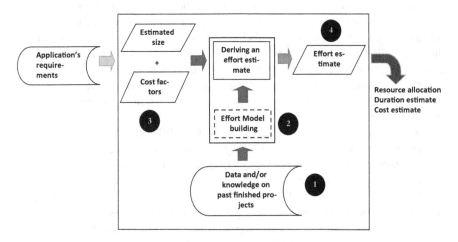

Fig. 3.9 Using CART for effort estimation

necessarily the case that all the initial independent variables are used to build a CART model; rather, only those variables that are related to the dependent variable are selected by the model. This means that a CART model can be used not only to produce a model that can be applicable for effort prediction, but also to obtain insight and understanding of the factors relevant to estimate a given dependent variable.

The sequence of steps (Fig. 3.9) followed here are as follows:

Step 1. Past data is used to generate a CART model.

Step 2. A CART model is built based on the past data obtained in Step 1.

Step 3. The model created in Step 2 then receives, as input, values/categories for the estimated size and cost drivers relative to the new project to which effort is to be estimated.

Step 4. The model generates a value/category for estimated effort.

The sequence of steps described above corresponds to the same sequence used with the algorithmic techniques. In both situations, data is used to either build an equation that represents an effort model, or to build a binary tree, which is later used to obtain effort estimates for new projects. This sequence of steps contrasts to the different sequence of steps used for expert opinion and CBR, where knowledge about a new project is used to select similar projects.

What Technique to Employ?

This chapter has introduced numerous techniques for obtaining effort estimates for a new project. These techniques were selected because they are the ones used the most in the Web effort estimation literature [32, 33], and to some extent also in the software effort estimation literature, each with a varying degree of success. Therefore the question that is often asked is: Which of the techniques provides the most accurate prediction for Web effort estimation?

To date, the answer to this question has been simply "it depends" [33–35].

Algorithmic and CART models have some advantages over CBR and expert opinion, such as:

- They allow users to see how a model derives its conclusions, an important factor for verification as well as theory building and understanding of the process being modelled [16].
- They often need to be specialised relative to the local environment in which they are used [36]. This means that the estimations that are obtained take full advantage of using models that have been calibrated to local circumstances.

Despite these advantages, no convergence for which effort estimation technique has the best predictive power has yet been reached, even though comparative studies have been carried out for nearly 20 years (e.g., [16, 17, 20–24, 26, 28, 30, 35, 37–42]).

One justification is that these studies often use data sets with differing number of characteristics (e.g., number of outliers, amount of collinearity, number of variables and number of projects) and different comparative designs. Note, an outlier is a value which is far from the others, and collinearity represents the existence of a linear relationship between two or more independent variables.

Shepperd and Kadoda [2] presented evidence showing there is a relationship between the success of a particular technique and factors such as training set size (size of the subset used to derive a model), nature of the "effort estimation" function (e.g., continuous or discontinuous) and characteristics of the data set. They concluded that the "best" prediction technique that can work on any type of data set may be impossible to obtain. Note, a continuous function is one in which "small changes in the input produce small changes in the output" (http://e.wikipedia.org/wiki/Continuous_function). If small changes in the input "can produce a broken jump in the changes of the output, the function is said to be discontinuous (or to have a discontinuity)" (http://e.wikipedia.org/wiki/Continuous_function).

Stepwise regression, CART and CBR have been applied to Web effort estimation, and have also had their prediction accuracy compared (Mendes et al. [22]). Mendes et al. [22] showed that stepwise regression provided the best results overall, and this trend has also been confirmed using a different data set of Web projects. However, the data sets employed in both studies presented a very strong linear relationship between size and cost drivers with effort, and as such it comes as no surprise that, of the three techniques, stepwise regression presented the best prediction accuracy. A more recent study presented contradictory findings, where in some situations Bayesian networks presented superior predictions [42].

What Datasets of Past Project Data to Use?

Most research in effort estimation has to date focused on solving companies' inaccurate effort predictions via investigating techniques that are used to build formal effort estimation models, in the hope that such formalization will improve the accuracy of estimates [33]. They do so by assessing, and often also comparing, the prediction accuracy obtained from applying numerous statistical and artificial intelligence techniques to datasets of completed industrial software/Web projects. Recent literature reviews of software and Web effort estimation studies are given in [43] and in [33], respectively.

Except for expert opinion, all the techniques detailed in this chapter require the use of a dataset containing past data on finished projects in order to provide an effort estimate for a new project. In regard to the type of dataset employed in previous studies, results suggest that datasets built using data from a single company, in particular when the data characterises Web projects, are more likely to provide superior estimation accuracy when compared to datasets that contain project data volunteered by different companies [44].

However, there are several issues that a company faces that are associated with building its own dataset of past projects, such as [44]:

- The time required to accumulate enough data on past projects from a single company may be prohibitive.
- By the time the dataset is large enough to be of use, technologies employed by the company may have changed, and older projects may no longer be representative of current practices.
- Care is necessary as data needs to be collected in a consistent manner.

Such issues have motivated previous studies to investigate to what extent effort estimation models built using cross-company (CC) datasets, i.e., datasets that contain project data coming from several different companies, can provide more accurate effort estimates for projects belonging to another company, when compared to effort estimates obtained using that company's own data (single-company (SC) dataset) [44]. To date 17 studies have looked at this issue, where only five used datasets of Web projects, and of these just three were independent studies. No overall consensus has been reached as to the merits of using CC datasets.

Practical Implications

The variables characterizing the types of datasets discussed above are determined in different ways, such as via surveys [3], interviews with experts [28], expertise from companies [45], a combination of research findings [25] or even a researcher's own consulting experience [46]. In all of these instances, once variables are defined, a data-gathering exercise takes place, obtaining data (ideally) from industrial projects volunteered by companies. Except when using research findings to inform variables' identification, invariably the mechanism employed to determine variables relies on experts' recalling, where the recalling time given to experts is short, and the subjective measure of an expert's certainty is often their amount of experience estimating effort. In other words, experts are expected to recall quickly and with a high level of certainty, without any checks and balances to validate their answer.

However, in addition to eliciting the size measures and cost factors believed to have an effect upon effort, such a mechanism does not provide the means to identify the cause and effect relationships between factors, and most of all, to quantify the uncertainty associated with these relationships and to validate the knowledge obtained. Why should these be important?

Our experience developing and validating several single-company expert-based Web effort prediction models that use a knowledge management methodology to incorporate the uncertainty inherent in this domain [47] showed that the use of a structured iterative process in which factors and relationships are identified, quantified and validated [48–51] leads the participating companies to a much more thorough and deep understanding of their mental processes and their decisions when estimating effort, when compared to just recalling factors and their relationships. The iterative process we use employs Bayesian inference, which is one of the techniques employed in root cause analysis [17]; therefore it aims at a detailed analysis and understanding of a particular phenomenon of interest.

In all the case studies we conducted, the original set of factors and relationships initially elicited was always modified as the model evolved; this occurred as a result of applying a root cause analysis approach comprising a Bayesian inference mechanism and feedback into the analysis process via a model validation. In addition, post-mortem interviews with the participating companies showed that the understanding companies gained by being actively engaged in building those models led to both improved estimates and estimation processes [48–51].

The next chapter will detail the technique that we have employed when building the expert-based Web effort prediction models, such that other companies can also employ this same technique as means to improve their current effort estimates and understanding regarding their effort estimation processes.

Conclusions

Effort estimation enables companies to know beforehand and before implementing an application, the amount of effort required to develop the application on time and within budget. To estimate effort, it is necessary to understand the project variables that may affect effort prediction and how they are interrelated. These variables represent an application's size (e.g., *number of New web pages and Images, the number of functions/features* (e.g., shopping cart) *to be offered by the new Web application*) and also include other factors that may contribute to effort (e.g., *total number of developers who will help develop the new Web application, developers' average number of years of experience with the development tools employed, main programming language used*).

The mechanisms used to obtain an effort estimate are generally classified as:

Expert-Based Estimation Expert-based effort estimation represents the process of estimating effort by subjective means, and is often based on previous experience with developing and/or managing similar projects. This is by far the most widely used technique for Web effort estimation.

Algorithmic-Based Estimation Algorithmic-based effort estimation attempts to build models (equations) that precisely represent the relationship between effort and one or more project characteristics via the use of algorithmic models (statistical methods that are used to build equations). These techniques have been to date the most popular techniques used in the Web and software effort estimation literature.

Estimation Using Artificial Intelligence Techniques Finally, artificial intelligence (AI) techniques are also used to obtain effort estimates, although not necessarily via a model, such as the ones created with algorithmic-based techniques. AI techniques include fuzzy logic [13], regression trees [14], neural networks [15] and case-based reasoning [2].

This chapter has detailed the use of case-based reasoning (CBR) and regression trees (CART), the two AI techniques that have been employed in the literature for Web effort estimation, and has also provided a discussion about the practical implications of using such techniques and projects' datasets.

References

1. Kitchenham BA, Pickard LM, Linkman S, Jones P (2003) Modelling software bidding risks. IEEE Trans Softw Eng 29(6):542–554

2. Shepperd MJ, Kadoda G (2001) Using simulation to evaluate prediction techniques. In: Proceedings of the IEEE 7th international software metrics symposium, London, UK, pp 349–358
3. Mendes E, Mosley N, Counsell S (2005) Investigating web size metrics for early web cost estimation. J Syst Softw 77(2):157–172
4. Jørgensen M, Sjøberg D (2001) Impact of effort estimates on software project work. Inf Softw Technol 43:939–948
5. DeMarco T (1982) Controlling software projects: management, measurement and estimation. Yourdon, New York
6. Vliet HV (2000) Software engineering: principles and practice, 2nd edn. Wiley, New York
7. Mendes E (2007) Predicting web development effort using a Bayesian network. In: Proceedings of EASE'07, pp 83–93
8. Gray R, MacDonell SG, Shepperd MJ (1999) Factors systematically associated with errors in subjective estimates of software development effort: the stability of expert judgement. In: Proceedings of the 6th IEEE metrics symposium, Boca Raton, FL, pp 216–226
9. Boehm B (1981) Software engineering economics. Prentice-Hall, Englewood Cliffs
10. Boehm B (2000) COCOMO II model definition manual. Retrieved January 2006, from The University of Southern California, web site: http://sunset.usc.edu/research/COCOMOII/Docs/modelman.pdf
11. Bohem B, Abts C, Brown A, Chulani S, Clark B, Horowitz E, Madachy R, Reifer D, Steece B (2000) In: Boehm B (ed) Software cost estimation with Cocomo II. Pearson
12. Schofield C (1998) An empirical investigation into software estimation by analogy. Unpublished Doctoral Dissertation, Department of Computing, Bournemouth University
13. Kumar S, Krishna BA, Satsangi PS (1994) Fuzzy systems and neural networks in software engineering project management. J Appl Intell 4:31–52
14. Schroeder L, Sjoquist D, Stephan P (1986) Understanding regression analysis: an introductory guide, no. 57. Sage, Newbury Park
15. Shepperd MJ, Schofield C, Kitchenham B (1996) Effort estimation using analogy. In: Proceedings of ICSE-18, Berlin, pp 170–178
16. Gray AR, MacDonell SG (1997) A comparison of model building techniques to develop predictive equations for software metrics. Inf Softw Technol 39:425–437
17. Ammerman M (1998) The root cause analysis handbook: a simplified approach to identifying, correcting, and reporting workplace errors. CRC Press, New York
18. Selby RW, Porter AA (1998) Learning from examples: generation and evaluation of decision trees for software resource analysis. IEEE Trans Softw Eng 14:1743–1757
19. Mendes E, Counsell S, Mosley N (2000) Measurement and effort prediction of web applications. In: Proceedings of 2nd ICSE workshop on web engineering, Limerick, Ireland, pp 57–74, June 2000
20. Kadoda G, Cartwright M, Chen L, Shepperd MJ (2000) Experiences using case-based reasoning to predict software project effort. In: Proceedings of the EASE 2000 conference, Keele, UK
21. Briand LC, El-Emam K, Surmann D, Wieczorek I, Maxwell KD (1999) An assessment and comparison of common cost estimation modeling techniques. In: Proceedings of ICSE 1999, Los Angeles, CA, pp 313–322
22. Mendes E, Watson I, Triggs C, Mosley N, Counsell S (2002) A comparison of development effort estimation techniques for web hypermedia applications. In: Proceedings IEEE metrics symposium, Ottawa, Canada, pp 141–151, June 2002
23. Jeffery R, Ruhe M, Wieczorek I (2000) A comparative study of two software development cost modelling techniques using multi-organizational and company-specific data. Inf Softw Technol 42:1009–1016
24. Jeffery R, Ruhe M, Wieczorek I (2001) Using public domain metrics to estimate software development effort. In: Proceedings of the 7th IEEE metrics symposium, London, pp 16–27

25. Mendes E, Mosley N, Counsell S (2001) Web measures – estimating design and authoring effort. IEEE Multimed (Special Issue on Web Eng) 8(1):50–57
26. Mendes E, Mosley N, Counsell S (2003) Do adaptation rules improve web cost estimation?. In: Proceedings of the ACM hypertext conference 2003, Nottingham, UK, pp 173–183
27. Mendes E, Mosley N, Counsell S (2003) A replicated assessment of the use of adaptation rules to improve web cost estimation. In: Proceedings of the ACM and IEEE international symposium on empirical software engineering, Rome, Italy, pp 100–109
28. Ruhe M, Jeffery R, Wieczorek I (2003) Cost estimation for web applications. In: Proceedings of ICSE 2003, Portland, OR, pp 285–294
29. Shepperd MJ, Schofield C (1997) Estimating software project effort using analogies. IEEE Trans Softw Eng 23(11):736–743
30. Mendes E, Mosley N, Counsell S (2002) The application of case-based reasoning to early web project cost estimation. In: Proceedings of IEEE COMPSAC, pp 393–398
31. Brieman L, Friedman J, Olshen R, Stone C (1984) Classification and regression trees. Wadsworth, Belmont
32. Mendes E (2007) Cost estimation techniques for web project. IGI Global, Hershey, p 424. ISBN 978-1-59904-135-3
33. Azhar D, Mendes E, Riddle P (2012) A systematic review of web resource estimation. In: Proceedings of PROMISE, New York, NY, pp 49–58
34. Mendes E (2008) The use of Bayesian networks for web effort estimation: further investigation. In: Proceedings of ICWE'08, Yorktown Heights, NJ, pp 203–216
35. Mendes E (2007) A comparison of techniques for web effort estimation. In: Proceedings of the ACM/IEEE international symposium on empirical software engineering, Madrid, pp 334–343
36. Kok P, Kitchenham BA, Kirakowski J (1990) The MERMAID approach to software cost estimation. In: Proceedings of the ESPRIT annual conference, Brussels, pp 296–314
37. Briand LC, Langley T, Wieczorek I (2000) A replicated assessment and comparison of common software cost modeling techniques. In: Proceedings of ICSE 2000, Limerick, Ireland, pp 377–386
38. Finnie GR, Wittig GE, Desharnais J-M (1997) A comparison of software effort estimation techniques: using function points with neural networks. Case-based reasoning and regression models. J Syst Softw 39:281–289
39. Gray A, MacDonell S (1997) Applications of fuzzy logic to software metric models for development effort estimation. In: Proceedings of IEEE annual meeting of the North American fuzzy information processing society - NAFIPS, Syracuse, NY, pp 394–399
40. Hughes RT (1997) An empirical investigation into the estimation of software development effort. Unpublished Doctoral Dissertation, Department of Computing, University of Brighton
41. Kemerer CF (1987) An empirical validation of software cost estimation models. Commun ACM 30(5):416–429
42. Mendes E, Mosley N (2008) Bayesian network models for web effort prediction: a comparative study. Trans Softw Eng 34(6):723–737
43. Jorgensen M, Shepperd M (2007) A systematic review of software development cost estimation studies. IEEE Trans Softw Eng 33(1):33–53
44. Kitchenham BA, Mendes M, Travassos GH (2007) Cross versus within-company cost estimation studies: a systematic review. IEEE TSE 33(5):316–329
45. Ferrucci F, Gravino C, Di Martino S (2008) A case study using web objects and COSMIC for effort estimation of web applications. In: EUROMICRO-SEAA, Parma, pp 441–448
46. Reifer DJ (2000) Web development: estimating quick-to-market software. IEEE Softw 17(6): 57–64
47. Mendes E (2012) Using knowledge elicitation to improve web effort estimation: lessons from six industrial case studies. In: Proceedings of the international conference on software engineering (ICSE' 2012), track SE in Practice, pp 1112–1121

48. Mendes E (2011) Building a web effort estimation model through knowledge elicitation. In: Proceedings of the 13th international conference on enterprise information systems, pp 128–135
49. Mendes E (2011) Improving project management of healthcare projects through knowledge elicitation. In: Miranda IM, Cruz-Cunha MM (eds) Handbook of research on ICTs for healthcare and social services: developments and applications, IGI Global (Accepted for publication)
50. Mendes E (2011) Uncertainty-based software effort estimation, IFPUG book, IFPUG (Accepted for Publication
51. Mendes E, Pollino C, Mosley N (2009) Building an expert-based web effort estimation model using Bayesian networks. In: Proceedings of the EASE conference, pp 1–10

Introduction to Web Resource Estimation

4

Introduction

Effort estimation, the process by which effort is forecasted and used as basis to predict costs and to allocate resources effectively, is one of the main pillars of sound project management, given that its accuracy can affect significantly whether projects will be delivered on time and within budget [1]. However, because it is a complex domain where corresponding decisions and predictions require reasoning with uncertainty, there are countless examples of companies that underestimate effort. Jørgensen and Grimstad [2] reported that such estimation error can be 30–40 % on average, thus leading to serious project management problems.

Similarly to software effort estimation, most research in Web effort estimation has to date focused on solving companies' inaccurate effort predictions via investigating techniques that are used to build formal effort estimation models, in the hope that such formalization will improve the accuracy of estimates. They do so by assessing, and often also comparing, the prediction accuracy obtained from applying numerous statistical and artificial intelligence techniques to datasets of completed Web projects developed by industry, and sometimes also developed by students. Details relating to previous studies in Web effort estimation are given next.

Systematic Literature Review on Web Resource Estimation

To understand effort estimation for Web projects, previous studies have developed models that use as input factors such as the size of a Web application and cost drivers (e.g., tools, developer's quality, team size), and provide an effort estimate as output. The differences between these studies were the number and type of size measures used, choice of cost drivers, and occasionally the techniques employed to build resource estimation models.

Mendes and Counsell [3] were the first, back in 2000, to investigate this field by using machine-learning techniques with data from student-based Web projects, and

E. Mendes, *Practitioner's Knowledge Representation*, DOI 10.1007/978-3-642-54157-5_4,   55
© Springer-Verlag Berlin Heidelberg 2014

size measures harvested late in the project's life cycle. Mendes and collaborators also carried out a series of consecutive studies (e.g., [1, 4–6]) where models were built using multivariate regression and machine-learning techniques using data on industrial Web projects. They also proposed and validated size measures harvested early in the project's life cycle, and therefore that were better suited to resource estimation [6].

Other researchers have also investigated resource estimation for Web projects, and some examples follows. Reifer [7] proposed an extension of the COCOMO model, and a single size measure harvested late in the project's life cycle. None were validated empirically. This size measure was later used by Ruhe et al. [8], who further extended a software engineering hybrid estimation technique to Web projects, using a small data set of industrial projects, mixing expert judgement and multivariate regression. Later, Baresi et al. [9], and Mangia and Paiano [10] investigated effort estimation models and size measures for Web projects based on a specific Web development method, namely the W2000. Finally, Costagliola et al. [11] compared two types of Web-based size measures for effort estimation.

Given that Web development is a rapidly growing industry [12], it is important to obtain a detailed account of the state of the art in this field in order to inform interested practitioners and researchers. Motivated by such need, Azhar et al. [12] have recently carried out a systematic literature review of Web resource estimation.

This section therefore provides a detailed summary of the above-mentioned systematic literature review on Web resource estimation [12]. The large majority of studies reported in this review (85.7 %) focused solely on Web effort estimation; therefore we assume that the results presented here represent the state of the art in the field of Web effort estimation.

A systematic review is a method that enables the evaluation and interpretation of all accessible research relevant to a research question, subject matter or event of interest [13]. There are numerous motivations for carrying out a systematic litera- ture review, amongst which the most common are [13]:

- to review the existing evidence regarding a treatment of technology, for exam- ple, to review existing empirical evidence of the benefits and limitations of a specific Web development method;
- to identify gaps in the existing research that will lead to topics for further investigation; and
- to provide a context/framework so as to properly place new research activities. A systematic review generally comprises the following steps [13]:
- formulation of a focused review question;
- identification of the need for carrying out a systematic review;
- a comprehensive, exhaustive search and inclusion of primary studies;
- quality assessment of included studies;
- data extraction;
- summary and synthesis of study results (meta-analysis);
- interpretation of the results to determine their applicability; and
- report-writing.

Prior to the review, it is desirable to develop a protocol that specifies the plan that the systematic review will follow to identify, assess and collate evidence.

A well-formulated question generally has four parts [13], identified as PICO (population, intervention, comparison, outcome):

- the population (e.g., the disease group, or a spectrum of the healthy population);
- the study factor (e.g., the intervention, diagnostic test or exposure);
- the comparison intervention (if applicable);
- the outcome.

The question should be sufficiently broad to allow examination of variation in the study factor and across populations [13].

The research questions addressed in the systematic literature review by Azhar et al. were the following:

Question 1: What methods and techniques have been used for Web resource estimation?

Question 1a: What metrics have been used to measure estimation accuracy?

Question 1b: What (numerical) accuracy did these various methods/techniques achieve?

Question 2: What resource facets (e.g., effort, quality, size) have been investigated in research on Web resource estimation?

Question 2a: What resource predictors have been used in the estimation process?

Question 2b: At what stage are these resource predictors gathered?

Question 3: What are the characteristics (single/cross-company, student/industry projects) of the datasets used for Web resource estimation?

These research questions address three main areas relating to Web resource estimation research, as follows:

1. The techniques used for Web resource estimation.
2. The resource facets that have been investigated and the predictors considered.
3. The characteristics of the datasets used in the empirical research.

A total of 84 studies were selected, after employing 11 different databases/ search engines to search for related literature in Web resource estimation.

Tables 4.1, 4.2, and 4.3 provide summaries respectively for each of the main areas mentioned above. Please refer to Azhar et al. [12] for more detailed tables and results.

A range of techniques have been used, which include expert judgment, various algorithmic and machine learning techniques, as well as those that fall into more than one category. Estimation accuracy forms the basis for evaluating these techniques, and a number of numerical and graphical measures of accuracy were employed most of which using as basis the absolute residual [4].

Table 4.1 Estimation
method/technique used for
Web resource estimation

Estimation method/technique	%
Case-based reasoning (CBR)	34.5
Stepwise regression	34.5
Linear regression	23.8
Bayesian networks	10.7
Classification and regression trees (CART)	6.0
Support vector regression	6.0
Expert judgment	4.8
Web-COBRA	4.8
Custom	13.1
Mean estimation	20.2
Median estimation	22.6
Other	16.7
No estimation method/technique	6.0

Table 4.2 Resource facets and predictors investigated in the studies

Resource facet investigated	%	Resource predictors	%
Design	3.6	Size: Length	50.0
Quality	3.6	Size: Functionality	32.1
Maintenance	6.0	Size: Reusability	21.4
Size	1.2	Complexity	34.5
Cost/effort	85.7	Cost drivers	19.0
		Tukutuku[a]	32.1
		Other	4.8
		No predictors investigated	1.2

[a]The Tukutuku database is part of the Tukutuku project, which aimed to collect data from completed Web projects, to be used to develop Web cost estimation models and to benchmark productivity across and within Web Companies [6]

Table 4.3 Domain and
type of dataset used in the
studies

Domain	%	Type	%
Industry	69.0	Cross-company	53.4
Academia	33.3	Single-company	50.0
Not specified	1.2		

Results showed that within the domain of Web resource estimation, work has been done on effort/cost, design, quality, maintenance and size estimation, where the main focus has been on development effort/cost estimation with only 14.4 % of the primary studies centring on other resource facets. In addition, out of the three studies focusing on quality estimation and out of the five studies dealing with maintenance estimation, only one and two provided an accuracy assessment, respectively. Such lack of accuracy assessment limits the usefulness of these studies for practitioners looking to undertake quality or maintenance estimation.

Size measures have historically been considered as key predictors of effort. This still holds true with length, reusability and functionality size measures being seen in 69 % of the selected primary studies. In addition, given that size measures are included amongst the Tukutuku variables [6], results show that, except for one study, every primary study that investigates resource predictors considered size measures as predictors of resource estimation.

Most of the research done to date in Web resource estimation employed predictors that presented an association with the resource facet being estimated, without assuming that this association was of the type cause and effect. However, there were some exceptions (studies using Bayesian nets and Web-COBRA), which used predictors that had a cause and effect relationship with the resource facet being estimated. These predictors were usually expert-based.

The review also showed that industry datasets were more frequently used than academic datasets, where such industry datasets contained either data from a single company (single company dataset), or from numerous companies (cross-company dataset). Estimates from single company datasets appeared to be superior to those from cross-company datasets, which corresponds to findings from prior research that has been done on single versus cross-company estimates, in both Web and general software resource estimation [14]. Single company datasets are smaller than their cross-company counterparts, of which the Tukutuku database is the largest and most often used.

The results from the systematic review showed that several estimation techniques have been employed, with no single technique providing the best accuracy results overall. In addition, most work focused on development effort/cost estimation, with little done on areas such as quality or maintenance effort estimation.

The lack of consensus on the best Web resource estimation techniques could be due to a number of reasons, such as [12]: choice of resource predictors and accuracy measures, dataset characteristics, and type of cross-validation employed, to name a few.

Overall, none of the previous studies used data gathered by any form of root cause analysis mechanism to build their predictions models. The very few studies that represented in some way or the other the uncertainty inherent to effort estimation did not do so based on experts' tacit knowledge. These two points are two of the main drivers for the research detailed herein.

Conclusions

This chapter presented the main findings relating to Web effort estimation based on the results from a systematic literature review on Web resource estimation [12]. Overall results showed that previous techniques did not employ root cause analysis methods or tacit knowledge from experts. This represents a gap in the state of the art in this area, which is partially filled by the six different case studies that are described in this book.

References

1. Mendes E, Mosley N, Counsell S (2001) Web metrics - metrics for estimating effort to design and author Web applications. IEEE Multimed 8(1):50–57
2. Jorgensen M, Grimstad S (2009) Software development effort estimation: demystifying and improving expert estimation. In: Tveito A, Bruaset AM, Lysne O (eds) Simula research laboratory - by thinking constantly about it. Springer, Heidelberg, pp 381–404, Chap. 26. ISBN 978-3642011559
3. Mendes E, Counsell S (2000) Web development effort estimation using analogy. In: Proceedings of 2000 Australian software engineering conference, Washington, DC, pp 203–212
4. Mendes E, Kitchenham BA (2004) Further comparison of cross-company and within-company effort estimation models for web applications. In: Proceedings of IEEE metrics, pp 348–357
5. Mendes E, Mosley N, Counsell S (2003) Investigating early web size measures for web cost estimation. In: Proceedings 7th international conference evaluation and assessment in software engineering (EASE 2003), pp 1–10, Apr 2003
6. Mendes E, Mosley N, Counsell S (2005) Investigating web size metrics for early web cost estimation. J Syst Softw 77(2):157–172. doi:10.1016/j.jss.2004.08.034
7. Reifer DJ (2000) Web development: estimating quick-to-market software. IEEE Softw 17(6): 57–64
8. Ruhe M, Jeffery R, Wieczorek I (2003) Cost estimation for web applications. Proc ICSE 2003: 285–294
9. Baresi L, Morasca S, Paolini P (2002) An empirical study on the design effort for web applications. Proc WISE 2002:345–354
10. Mangia L, Paiano R (2003) MMWA: a software sizing model for web applications. In: Proceedings of fourth international conference on web information systems engineering, pp 53–63
11. Costagliola G, Di Martino S, Ferrucci F, Gravino C, Tortora G, Vitiello G (2006) Effort estimation modeling techniques: a case study for web applications. In: Proceedings of international conference on web, engineering (ICWE'06), New York, NY, pp 9–16
12. Azhar D, Mendes E, Riddle P (2012) A systematic review of web resource estimation. In: Proceedings of promise'12, New York, NY, pp 49–58
13. Kitchenham BA (2007) Guidelines for performing systematic literature reviews in software engineering (version 2.3). Software Engineering Group, School of Computer Science and Mathematics, Keele University; Department of Computer Science, University of Durham, Jul 2007
14. Kitchenham B, Mendes E, Travassos GH (2006) A systematic review of cross-company vs. within-company cost estimation studies. In: Proceedings 10th international conference on evaluation and assessment in software engineering (EASE 2006), pp 89–98

Introduction to Bayesian Networks

5

Introduction

There are numerous knowledge management techniques available (e.g., [1, 2]) that can be used by practitioners to support decision making where reasoning takes place under uncertainty. This chapter presents one of these knowledge management techniques, which is also the technique employed in all the six case studies detailed in Chaps. 7–12.

Note that within the context of this book reasoning under uncertainty means that there is no deterministic solution to a decision being discussed. This is a typical situation in complex domains such as effort estimation. For example, assuming there is a relationship between an application's development effort and this application's size (e.g., number of Web pages, functionality), it is not necessarily true that an increase in an application's size will always lead to an increase in this application's development effort. However, as an application's size grows there is the *probability* that its development effort is also likely to increase. This therefore means that any decisions relating to a development effort estimate for an application where this application's size has been identified as large will likely also lead to a high development effort.

The technique described herein is called Bayesian network (BN) [3]. Bayesian networks have been successfully employed for decision-making under uncertainty in several complex domains (e.g., genetics, speech recognition, medical diagnosis, software project management) [4]. This technique was chosen within the context of this book for three main reasons:

1. It supports reasoning under uncertainty given the way it incorporates knowledge of a complex domain [3].
2. It incorporates three of the four stages of a knowledge creation process [1, 5].

E. Mendes, *Practitioner's Knowledge Representation*, DOI 10.1007/978-3-642-54157-5_5,   61
© Springer-Verlag Berlin Heidelberg 2014

3. We have previously applied this technique successfully to support decision-making under uncertainty in three complex domains—software resource estimation, software risk management and software requirements prediction—all collaborations with numerous industry partners worldwide [4, 6–9].[1]

Bayesian Network

A Bayesian network (BN) is a technique that enables the construction of a model (BN model) that supports reasoning with uncertainty due to the way in which it incorporates existing knowledge of a complex domain [3]. Its main components will be explained using an example model that is first shown and detailed in Fig. 5.1 and in Table 5.1, respectively. This is the same example model used in Chaps. 1–6, and it is meant to be an example of a BN model for estimating development effort for Web applications.

Knowledge is represented in a BN model using two parts. The first, the qualitative part, represents the structure of a BN model (Fig. 5.1). This structure includes the relevant factors in the domain being modelled (e.g., project planning overhead, average team's expertise) and their causal relationships [3]. A BN model's structure can be elicited from experts, learned from data, or created using a combination of both. Within the context of this book, all structures were elicited from experts.

The second, quantitative part, quantifies probabilistically all the causal relationships that were identified in the qualitative part, using tables called conditional probability tables (CPTs). Each table describes a probability distribution. Similarly to a BN's structure, CPTs can be elicited from experts, learned from data, or created using a combination of both. Within the context of this book, all CPTs were elicited from experts.

Figures 5.2, 5.3 and 5.5 show the CPTs for factors "Average Team's Expertise", "Technological diversity", and "Combined Cost Factors' Effort", respectively.

Figure 5.2 shows the conditional probability table for factor "Average Team's Expertise". This factor is measured using five categories (further details about this are given in Chap. 6)—very high, high, average, low and very low. Each category has a number underneath, all adding to 100 [3]. Within the context of this example, each of these numbers represents the frequency of occurrence of a given category, given a set of past projects. Such sets of past projects may represent projects that were managed over the past year, over the past 3 years, all the projects managed by a company and so on. Figure 5.2 shows that 35 % of past projects presented very high "Average Team's Expertise"; 25 % of past projects presented high "Average Team's Expertise"; 25 % of past projects presented average "Average Team's

---

[1] This work was funded by the Royal Society of New Zealand (Marsden Fast Start research grant 06-UOA-201), and a Research Fellowship by the Brazilian Agency for Scientific Improvement. Professor Mendes was the first female in CS in NZ to obtain a Marsden FS as sole investigator.

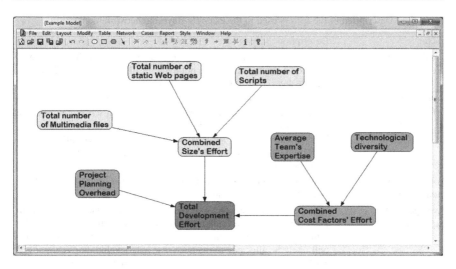

Fig. 5.1 Example Bayesian network model

Expertise"; 10 % of past projects presented low "average team's expertise"; and 5 % of past projects presented very low "Average Team's Expertise".

Figure 5.3 shows the conditional probability table for another factor—"Technological diversity". This factor is also being measured using five categories (further details about this are given in Chap. 6)—Very High, High, Average, Low and Very Low. Each category has a number underneath, all adding to 100 [3]. Within the context of this example, each of these numbers represents the frequency of occurrence of a given category given a set of past projects. Figure 5.3 shows that 35 % of past projects presented very high "Technological diversity"; 30 % of past projects presented high "Technological diversity"; 20 % of past projects presented average "Technological diversity"; 10 % of past projects presented low "Technological diversity"; and 5 % of past projects presented very low "Technological diversity".

These two CPTs are relatively easy to quantify as their probabilities do not depend on any existing causal relationships with other factors part of the same BN model. However, the quantification of categories for the next CPT is much more challenging given that factor "Combined Cost Factors' Effort" is affected by two other factors (causes): "Average Team's Expertise" and "Technological diversity" (Fig. 5.4). Note that this factor is also characterised using five categories (very high, high, average, low, very low).

As a result, all the quantifications that are provided for the categories characterising factor "Combined Cost Factors' Effort" need to take into account the categories for factors "Average Team's Expertise" and "Technological diversity" using decision scenarios. Two of these decision scenarios are highlighted in red in Fig. 5.5. The first one relates to a decision where past projects presented very high "Technological diversity" and very high "Average Team's Expertise". In such scenario, there is a probability of 75 % that "Combined Cost Factors' Effort" will be

Table 5.1 Description of the factors used in the example Bayesian network model

Factor	Description
Total number of static Web pages	This factor represents the estimated number of new Web pages that need to be created. These are not dynamically-generated pages, and include any types of page such as .htm, .html, .php
Total number of scripts	This factor represents the estimated sum of any types of scripts that are likely to be created for the Web application. Such scripts can be made using a client-side scripting technique (e.g., xml, Ajax techniques, Flash ActionScript), or server-side scripting languages (ASP, JSP, Perl, PHP, Python). It also includes files written using Cascading Style Sheets (CSS)
Total number of multimedia files	This factor represents the total estimated number of any multimedia content, such as images and videos
Average team's expertise	This factor measures team expertise as the average number of years of experience that the development team has with Web development. The estimation within this context relates to tentative decisions as to who will likely be allocated to the team that will develop the Web application for which total effort is being estimated
Technological diversity	This factor represents the estimated amount of diversity as far as the use of technology is concerned. It is measured using a surrogate measure, represented by the number of different technologies that are being employed in order to develop a Web application. Examples of technologies are MySQL, PHP, HTML, CSS, Python, ASP and JSP
Project planning overhead	This factor represents the degree of participation needed by the project manager in order to ensure the project is managed adequately and is ideally completed within time and on budget. This includes, but is not limited to status reports; communication; implementation plan (more for large projects) which includes the tasks to be done and their estimated completion dates; risk analysis; data analysis; planning (project execution plan)
Total development effort	This factor represents the total estimated effort to develop a Web application. The three factors that have a direct effect upon total effort are: combined cost factors' effort, project planning overhead and combined size's effort
Combined size's effort	This factor represents the estimated amount of effort (person-hours) needed to create Web pages, scripts/CSS files and multimedia files. Note that the effort will change depending on which categories are selected for each factor. Such selection will take place as part of a decision making scenario, and examples will be given later on (See Figs. 5.6 and 5.7)
Combined cost factors' effort	This factor represents the estimated amount of effort (person-hours) when taking into account technological diversity and average team's expertise. Note that the effort will change depending on which categories are selected for each factor. As previously stated, such selection will take place as part of a decision making scenario, and examples will be given later on

very low, and a 25 % probability that "Combined Cost Factors' Effort" will be low. This quantification should reflect experts' past experience with that specific decision scenario, taking into account the set of projects that are being considered

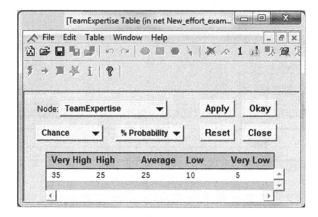

Fig. 5.2 CPT for factor "team's expertise"

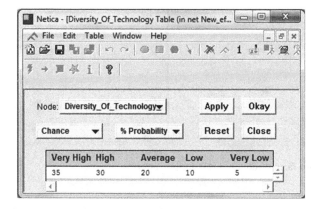

Fig. 5.3 CPT for factor "technological diversity"

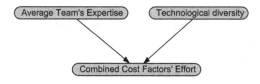

Fig. 5.4 Part of the BN model's structure showing three factors

during the quantification (e.g., all projects, projects completed over the past year, etc.). The second scenario relates to a decision where past projects presented very high "Technological diversity" and low "Average Team's Expertise". In such scenario, there is a probability of 85 % that "Combined Cost Factors' Effort" will be very high, and a 15 % probability that "Combined Cost Factors' Effort" will be

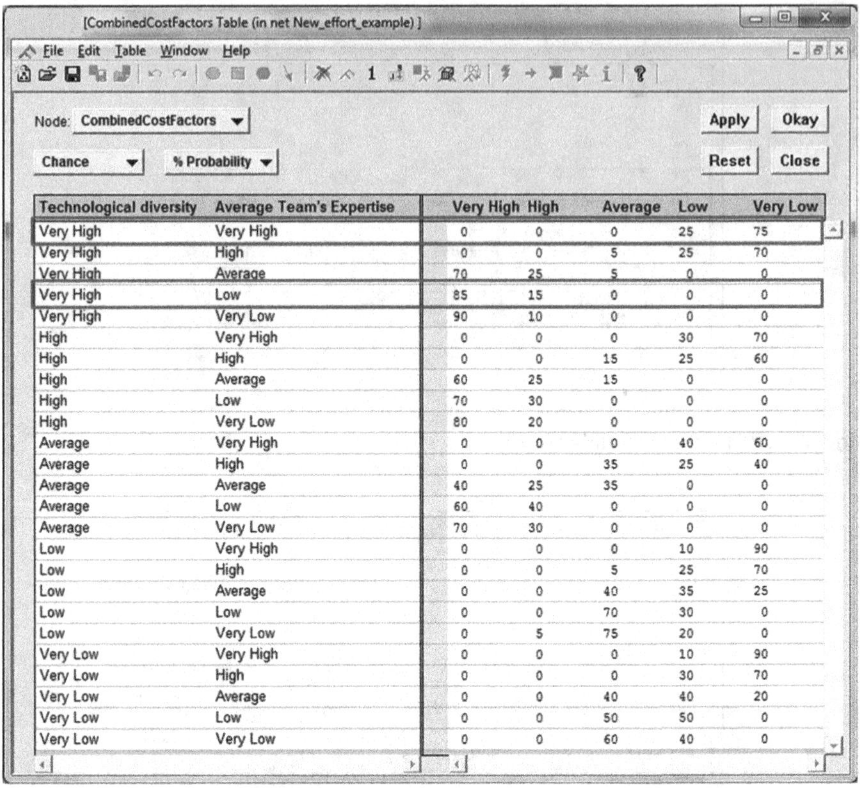

[CombinedCostFactors Table (in net New_effort_example)]

File   Edit   Table   Window   Help

Node: CombinedCostFactors ▼   Apply   Okay

Chance ▼   % Probability ▼   Reset   Close

Technological diversity	Average Team's Expertise	Very High	High	Average	Low	Very Low
Very High	Very High	0	0	0	25	75
Very High	High	0	0	5	25	70
Very High	Average	70	25	5	0	0
Very High	Low	85	15	0	0	0
Very High	Very Low	90	10	0	0	0
High	Very High	0	0	0	30	70
High	High	0	0	15	25	60
High	Average	60	25	15	0	0
High	Low	70	30	0	0	0
High	Very Low	80	20	0	0	0
Average	Very High	0	0	0	40	60
Average	High	0	0	35	25	40
Average	Average	40	25	35	0	0
Average	Low	60	40	0	0	0
Average	Very Low	70	30	0	0	0
Low	Very High	0	0	0	10	90
Low	High	0	0	5	25	70
Low	Average	0	0	40	35	25
Low	Low	0	0	70	30	0
Low	Very Low	0	5	75	20	0
Very Low	Very High	0	0	0	10	90
Very Low	High	0	0	0	30	70
Very Low	Average	0	0	40	40	20
Very Low	Low	0	0	50	50	0
Very Low	Very Low	0	0	60	40	0

Fig. 5.5 CPT for factor "combined cost factors' effort"

high. Each scenario represents a combination of every category for every factor that affects the target factor (in this example, "Combined Cost Factors' Effort"). In summary, the CPT "Combined Cost Factors' Effort" describes the relative probability of each category, conditional on every combination of categories of its parents (factors that affect "Combined Cost Factors' Effort" directly).

The outcome from applying this technique is the creation of a BN model (Fig. 5.6), which can then be used to run "what-if" scenarios that are used for reasoning and decision-making under uncertainty. One example of a what-if scenario is shown in Fig. 5.7. This scenario shows a change in factor "Project Planning Overhead", where from a situation where there was a probability distribution associated with all of its categories, it changes into a state where a probability of 100 % is associated with its category "Very Low" (Fig. 5.8). Such a change where a single category is associated with a 100 % probability is called to enter evidence, i.e., data about a factor (where data is represented in a categorical form) that has become known to (has been observed by) those who are using the model. Once such evidence is entered in the model, it triggers a change in the probabilities for all the

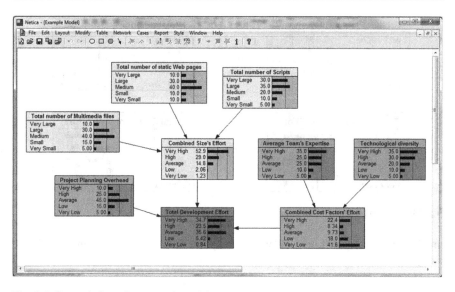

Fig. 5.6 Example Bayesian network model

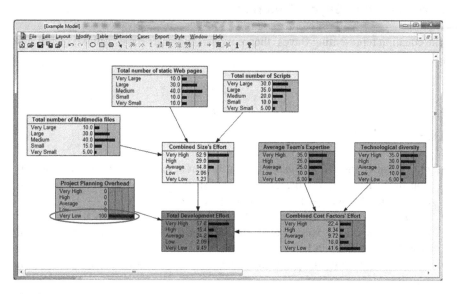

Fig. 5.7 "What-if" scenario

categories categorising factor "Total Development Effort". This is the outcome from applying a theorem that describes the relationship between the probability of a hypothesis given some evidence and the probability of that evidence given the hypothesis [3], i.e., a theorem that defines the relation between a conditional probability and its inverse form. This theorem is called Bayes' theorem [3], and a

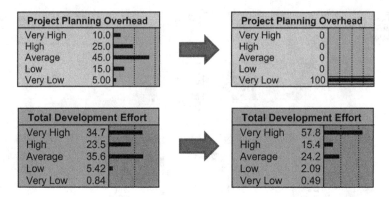

Fig. 5.8 Entering evidence and updated probabilities in the target factor

small example is given next in order to explain how probabilities are calculated using it.

Understanding Bayes' Theorem

The equation representing Bayes' theorem (also known as Bayes' rule) is as follows [3]:

$$p(X|Y) = \frac{p(Y|X)p(X)}{p(Y)} \qquad (5.1)$$

where:

- $p(X)$ is called the *prior* probability distribution or *marginal* probability distribution of X. It represents the probability distribution for X without taking into account any knowledge relating to the evidence Y.
- $p(X|Y)$ is called the *posterior* probability distribution of X as its values are dependent upon the value given for evidence Y. It represents the conditional probability of X given Y.
- $p(Y|X)$ is the conditional probability of Y given X. It is also called the *likelihood* function.
- $p(Y)$ is the *prior* or *marginal* probability distribution of evidence Y, and is used as a normalising constant.

Let's look at a very simple BN model and corresponding CPTs (Fig. 5.9) to understand how Bayes' theorem works.

This model shows a causal relationship between "Application size" and "Development Effort". The CPT for "Application size" shows that 40 % of past projects presented a large application size, and 60 % a small size. The CPT for "Development Effort" shows that if "Application size" is large, there is a 90 % chance that development effort will be high, and a 10 % chance that development effort will be

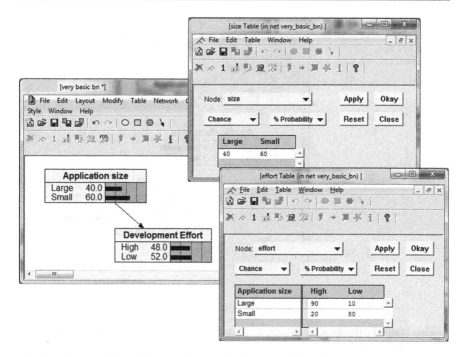

Fig. 5.9 Very small Bayesian network model and corresponding CPTs

low. On the other hand, if "Application size" is small there is a 20 % chance that development effort will be high, and an 80 % chance that development effort will be low.

Now let's focus upon two events:

Application size being large—event X

Development effort being high—event Y

Given these two events, we can assert the following:

- $p(X) = 0.40 = $ Probability of application size being large.
- $p(\sim X) = 0.60 = $ Probability of application size being small.
- $p(Y|X) = 0.90 = $ Probability of development effort being high, if application size is large.
- $p(\sim Y|X) = 0.10 = $ Probability of development effort being low, if application size is large.
- $p(Y|\sim X) = 0.20 = $ Probability of development effort being high, if application size is small.
- $p(\sim Y|\sim X) = 0.80 = $ Probability of development effort being low, if application size is small.

Given such scenarios, let's first determine the probability that development effort is high. This is obtained by adding the probability of application size being large and development effort being high to the probability of application size being small and development effort being high, as follows:

Fig. 5.10 Probability that
application size is large given
the evidence that
development effort is high

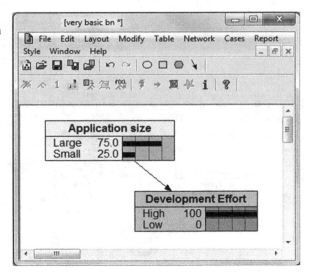

$$(40.90) + (60.20) = 3,600 + 1,200 = 4,800 = 48\,\%$$

This result is shown in Fig. 5.9; i.e., there is a 48 % chance that development effort will be high.

Let's now look at another scenario, in which development effort is known to be high. Given this evidence, what is the probability that application size will be large? The outcome is shown in Fig. 5.10; however, we will use Bayes' theorem to show how the probability was obtained. Bayes' theorem enables us to compute the probability of an earlier event given existing knowledge (evidence) about a later event.

We have already shown Bayes' theorem as Eq. (5.1); X represents the event application size being large, and Y represents the event development effort being high. Therefore, we have:

- $P(Y|X) = 90 =$ Probability of development effort being high, if application size is large.
- $P(X) = 40 =$ Probability of application size being large.
- $P(Y) = 48 =$ Probability of development effort being high.

Therefore, to compute the probability of application size being large given that development effort is high, we have the following equation:

$$p(X|Y) = \frac{90 \cdot 40}{48} \tag{5.2}$$

Thus, this probability is equal to 75 (75 %), as shown in Fig. 5.10.

In summary, once a BN model is specified, evidence (e.g., values for categories) can be entered into any factor, and probabilities for the remaining factors are automatically calculated using Bayes' theorem [3]. Therefore BN models can be used for different types of reasoning, such as predictive, diagnostic, in combination with what-if scenarios or to investigate the impact that changes on some factors have upon others.

Note that without tool support it is not possible to employ such models for decision making.

Conclusions

This chapter has provided an introduction to Bayesian network models by detailing and explaining their two main parts—qualitative and quantitative.

It also provided an introduction to Bayes' theorem and two simple examples so to provide the reader with a more detailed understanding of what happens when we use a Bayesian network model for decision making under uncertainty.

References

1. Darwiche A (2010) Bayesian networks. Commun ACM 53:80–90
2. Lempert R, Nakicenovic N, Sarewitz D, Schlesinger M (2004) Characterizing climate-change uncertainties for decision-makers. An editorial essay. Clim Chang 65:1–9
3. Pearl J (1988) Probabilistic reasoning in intelligent systems: networks of plausible inference. Morgan Kaufmann, San Francisco
4. Mendes, E. (2012) Using knowledge elicitation to improve Web effort estimation: Lessons from six industrial case studies. In: Practice in international conference on software engineering (accepted for publication), Zurich
5. Nonaka I, Toyama R (2003) The knowledge-creating theory revisited: knowledge creation as a synthesizing process. Knowl Manag Res Pract 1:2–10
6. Mendes E (2009) Using bayesian networks for web effort estimation. In: Meziane F, Vadera S (eds) Artificial intelligence applications for improved software engineering development: new prospects. IGI Global, Hershey, pp 26–44
7. Mendes E (2011) Knowledge representation using Bayesian networks; a case study in Web effort estimation. In: World congress on information and communication technologies (WICT), Mumbai, pp 612–617
8. Mendes E, Pollino C, Mosley N (2009) Building an expert-based Web effort estimation model using Bayesian networks. In: 13th international conference on evaluation and assessment in software engineering
9. Mendes E (2011) Building a Web effort estimation model through knowledge elicitation. In: Proceedings of the international conference on enterprise information systems (ICEIS)

Expert-Based Knowledge Engineering of Bayesian Networks

6

Introduction

The previous chapter provided an introduction to Bayesian network models, which are models that can be built solely from expert domain knowledge, solely from data, or by employing a combination of knowledge and data (hybrid models). However, within the context of Web effort estimation there are issues with building data-driven or hybrid Bayesian network models, as follows:

1. Any dataset used to build a Bayesian network model should be large enough to provide sufficient data capturing all (or most) relevant combinations of values for all the categories characterising factors such that probabilities can be learned from data, rather than elicited manually. Under such circumstance, it is very unlikely that the dataset would contain project data volunteered by only a single company (single-company dataset). As far as we know, the largest dataset of Web projects available is the Tukutuku dataset (195 projects) [1]. This dataset has been used to build data-driven and hybrid Bayesian network models; however, results have not been encouraging overall, and we believe one of the reasons is due to the small size of this dataset.

2. Even when a large dataset is available, the next issue relates to the set of factors part of the dataset. It is unlikely that the factors identified represent all those within a given domain (e.g., Web effort estimation) that are important for companies that use the data-driven or hybrid model created from this dataset. This was the case with the Tukutuku dataset, even though the selection of which factors to use had been informed by two surveys [1]. However, one could argue that if the model being created is hybrid, then new factors can be added to, and existing factors can be removed from the model. The problem is that every new factor added to the model represents a set of probabilities that need to be elicited from scratch, which may be a hugely time-consuming task.

3. Different structure and probability learning algorithms can lead to different prediction accuracy [2]; therefore one may need to use different models and

E. Mendes, *Practitioner's Knowledge Representation*, DOI 10.1007/978-3-642-54157-5_6,   73
© Springer-Verlag Berlin Heidelberg 2014

compare their accuracy, which may also be a very time-consuming task, in particular for companies willing to use such models.

4. When using a hybrid model, the Bayesian network's structure should ideally be jointly elicited by more than one domain expert, preferably from more than one company; otherwise the model built may not be general enough to cater for a wide range of companies [2]. There are situations, however, where it is not feasible to have several experts from different companies cooperatively working on a single Bayesian network structure. One such situation is when the companies involved are all consulting companies potentially sharing the same market. This was the case within the context of the research that motivated the writing of this book.

5. Ideally the probabilities used by the data-driven or hybrid Bayesian network models should be revisited by at least one domain expert, once they have been automatically learned using the learning algorithms available in Bayesian network tools. However, depending on the complexity of the Bayesian network model, this may represent having to check thousands of probabilities, which may not be feasible. One way to alleviate this problem is to add additional factors to the Bayesian network model in order to reduce the number of causal relationships reaching factors; however, all probabilities for the additional factors would still need to be elicited from domain experts.

6. The choice of factor discretisation, structure learning algorithms, probability estimation algorithms and the number of categories used in the discretisation all affect the accuracy of the results, and there are no clear-cut guidelines on what would be the best choice to employ. It may simply be dependent on the dataset being used, the amount of data available and trial and error to find the best solution [2].

Therefore, given the above-mentioned constraints, this chapter details a process that targets building expert-based Bayesian network models. This is also the process employed when building all the models that are detailed in Chaps. 7–12.

Note that we are not suggesting that data-driven and hybrid Bayesian network models should not be used. On the contrary, they have been successfully employed in numerous domains [3–5]; however, the specific domain context of this paper— that of Web effort estimation—provides other challenges (described above) that lead to the development of solely expert-driven Bayesian network models.

Introducing the Expert-Based Knowledge Engineering of Bayesian Networks Process

The process that we have followed in order to build all the expert-based Bayesian network models detailed in Chaps. 7–12 is shown in Fig. 6.1. It has been adapted from Woodberry et al. [3] in order to provide a set of steps that are achievable and meaningful to the practitioners participating in the process. In Fig. 6.1 arrows represent flows through the different subprocesses, which are depicted by

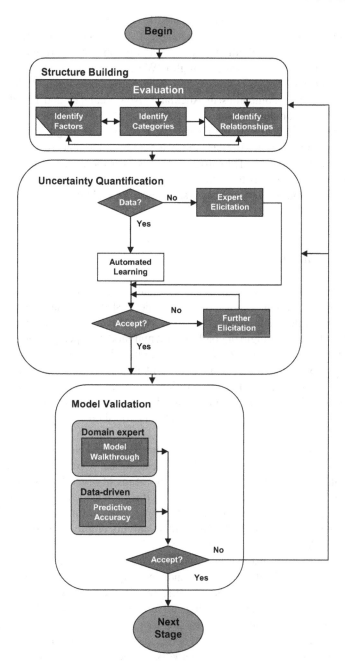

Fig. 6.1 Expert-based knowledge engineering of Bayesian networks

rectangles. These subprocesses are executed by people—the knowledge engineer and the domain experts (dark grey rectangles), by automatic algorithms (white rectangles), or via a combination of people and automatic algorithms (dark grey rectangle with bottom-left white triangle). Within the context of this book the author was the knowledge engineer, and Web project managers, Web designers and developers from companies in Auckland (New Zealand) and Rio de Janeiro (Brazil) were the domain experts.

The three main steps within the Expert-based Knowledge Engineering of Bayesian Networks process (EKEBN) are the structure building, uncertainty quantification, and model validation. This process iterates over these steps until a complete Bayesian network model is built and validated. Each of these three steps is briefly introduced next, followed by a step-by-step description of how to apply each one in practice.

Structure Building

The structure building step represents the qualitative component of a Bayesian network model, which results in a graphical structure that represents the explicitation of the tacit knowledge from one or more domain experts. This explicitation leads to a single or combined mental model (depending on the number of domain experts participating in the structure building step) that includes, in our case, the factors and causal relationships identified as fundamental for effort estimation of Web projects. In addition to identifying factors and causal relationships, this step also includes the decision of how each factor will be measured, i.e., each of the categories that a factor should take (e.g., very high, very low).

This step explicitates a mental model that is refined iteratively through the sub-processes "identify factors", "identify categories", "identify relationships", and evaluation. Such iteration takes place by means of several knowledge elicitation and representation meetings, attended by the knowledge engineer and domain expert(s). This type of structure construction process has been validated in previous studies [6] and uses the principles of problem solving employed in data modelling and software development [4]. As will be detailed later in this chapter, existing literature in Web effort estimation is also included as part of this step.

In addition, throughout this step the knowledge engineer evaluates the Bayesian network's structure in two stages. The first entails checking whether: factors and their categories have a clear meaning, all relevant factors have been included, factors are named conveniently, all categories are appropriate (exhaustive and exclusive), and a check for any categories that can be combined. The second stage entails reviewing the Bayesian network's structure in order to check whether the cause and effect relationships that were identified by the domain expert (s) genuinely correspond to their mental models. Once the Bayesian network's structure is assumed by the domain expert(s) to be close to final, the knowledge engineers may still need to optimise this structure to reduce the number of probabilities that need to be elicited. As will be detailed later on, if an optimisation is needed, the knowledge engineer and the domain expert(s) discuss possible choices and jointly propose new factors that are solely created in order to optimise the structure.

Note that it is important that two distinct roles are represented by those participating in the creation of a Bayesian network model—knowledge engineer and domain expert. We would suggest that someone who has expertise with eliciting requirements from clients to be chosen as the knowledge engineer, simply because they will have some experience in how to elicit tacit knowledge even if such knowledge has a different nature to the one being represented in the Bayesian network model. It is also important to stress herein that the knowledge engineer is a facilitator, and therefore they should be neutral and completely refrain from also taking the role of a domain expert.

Uncertainty Quantification

The uncertainty quantification step represents the quantitative component of a Bayesian network, where conditional probabilities corresponding to the quantification of the relationships between factors [7] are obtained. Such probabilities can be attained via expert elicitation, automatically from data, from existing literature or by using a combination of these. However, within the context of this book we will only discuss probabilities that are obtained via expert elicitation.

Model Validation

The model validation step, as the names implies, validates the Bayesian network model that results from the two previous steps, and determines whether it is necessary to revisit any of those steps. Two different validation methods are generally used, model walk-through and predictive accuracy. These two validation methods are briefly presented below, and will be detailed later in this chapter, as both were used by the author when building all the Bayesian network models that are detailed in Chaps. 7–12.

Model walk-through represents the use of real case scenarios that are prepared and used by the domain expert(s) to assess whether the predictions provided by a Bayesian network model presenting the highest probability correspond to the predictions that the domain expert(s) would have chosen based on their own expertise. Success is measured as the frequency with which the Bayesian network model's predictions correspond to the experts' own assessment.

Predictive accuracy uses data from past finished projects (which within our context are Web projects), rather than scenarios, to obtain predictions. Data (evidence) corresponding to categories from existing factors are entered on the Bayesian network model, and success is measured as the frequency with which the Bayesian network model's category for a target factor (e.g., effort) showing the highest probability corresponds to the actual past data.

Detailing the EKEBN Process

This section revisits the EKEBN steps discussed above and provides a detailed explanation of what to do in each of its steps. Although all the details and discussions presented herein will focus on the domain of effort estimation for Web projects, note that the methodology described relating to the building of Bayesian network models using the EKEBN process can equally be used when building models in any other domains (e.g., quality prediction, risk assessment, ecology or sustainability prediction).

Before detailing the EKEBN process, it is important to understand that the three steps that are part of that process all represent mechanisms used to achieve the externalisation, combination (optional) and internalisation of knowledge, as per Nonaka and Toyama's theory of organisational knowledge creation [8] (Fig. 6.2). As discussed in Chap. 1, this is the theory that has been adopted by the author within the context of this book, given that this theory was used as the basis when building all the expert-based Web effort estimation Bayesian network models that are detailed in Chaps. 7–12.

How Does the Externalisation of Knowledge Occur During the EKEBN Process?

Whenever we plan to build an expert-based Bayesian network model we need at least two people to take part in such a process, representing respectively the roles of knowledge engineer and domain expert. The knowledge engineer is responsible for eliciting the tacit knowledge from the domain expert. In addition, in order for this elicitation to be effective it is important that the domain expert explains as much as possible how they estimate effort for new projects. Of course, the knowledge engineer does not have a passive role during the elicitation meetings, as their goal is to actively facilitate the immersion of the domain expert(s) into the effort estimation process (which is generally done using decision making scenarios and case studies) such that all the fundamental factors, relationships and uncertainty quantifications are obtained. At the very start of the elicitation process, as will be detailed next, a whiteboard (or any other means that allows one to display the factors that are to be discussed) is used as a working surface on which factors, relationships and categories are recorded until all the domain experts participating in the elicitation process agree that an acceptable draft has been attained. Such factors and relationships represent explicit knowledge, and all the discussions that take place during elicitation meetings are focused on making as explicit as possible the domain expert(s)' tacit knowledge (knowledge explicitation/externalisation).

Fig. 6.2 The four different stages of the theory of organisational knowledge creation [8]. In this figure, I means Individual, G means Group, and O means Organisation

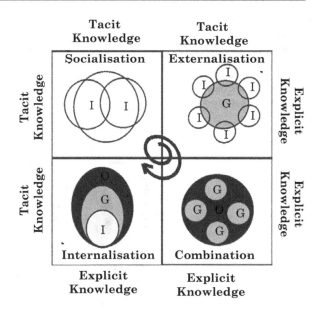

How Can the Combination of Knowledge Occur During the EKEBN Process?

All the Bayesian network models that are detailed in Chaps. 7–12 employ two of the knowledge creation steps, as per the theory of organisational knowledge creation—externalisation and internalisation of knowledge (Fig. 6.3). However, the combination of knowledge can also occur, as detailed next.

When it comes to combining different Bayesian network models, to date there is no universal automatic solution to merge diverse Bayesian network models, in particular if such models vary in the number of factors, corresponding categories and relationships. Some work has been done to create a semiautomatic approach to aggregating structures (factors and relationships) from different Bayesian network models [6]. However, this aggregation approach does not tackle the combination of probabilities and diverse categories, so further work needs to be done in order to aggregate complete Bayesian network models.

Given our experience building several expert-based Bayesian network models, we suggest that companies that are willing to also combine their models in a pragmatic and effective way should employ the following adaptation of the EKEBN process (Fig. 6.4), which is also represented in Fig. 6.5 using the theory of organisational knowledge creation.

Figure 6.4 shows an additional step called "structure combination". This step represents the merging of all the factors, and their corresponding categories and relationships that were previously identified by different groups in a company or organisation. The participants should include at least one domain expert from each group and also the knowledge engineer(s) who have participated in the structure

Fig. 6.3 The two different stages of the theory of organisational knowledge creation that were used when building expert-based Bayesian network models presented in Chaps. 7–12

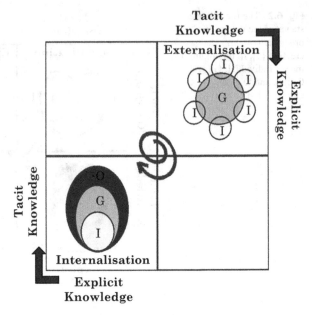

building step. The tangible outcome from this step should be a single Bayesian network model that embeds all the factors, categories and relationships that were identified as fundamental by the groups. It is important that agreement be reached by consensus so to guarantee that the aggregated model represents all the views from the different groups.

We suggest that the structure combination step be carried out prior to the uncertainty quantification step, and that both uncertainty quantification and model validation be conducted as a team effort with the participation of at least one domain expert from each group and also the knowledge engineer(s) who participated in the structure building and structure combination steps. This is suggested in order to prevent the need to also aggregate different conditional probability tables, which could lead to a major headache and the need to spend an enormous amount of time trying to reach consensus on every decision scenario.

How Does the Internalisation of Knowledge Occur During the EKEBN Process?

Whenever there are at least two domain experts they engage in very detailed discussions as the goal of all the elicitation meetings is to always reach a consensus amongst its members relating to factors, relationships, categories and, later on, also probabilities. As they discuss, and keeping in mind that from the start the knowledge engineer does not use a "blank slate", they use as basis the knowledge that has been explicitated. Such knowledge can be solely their own, or based on results from

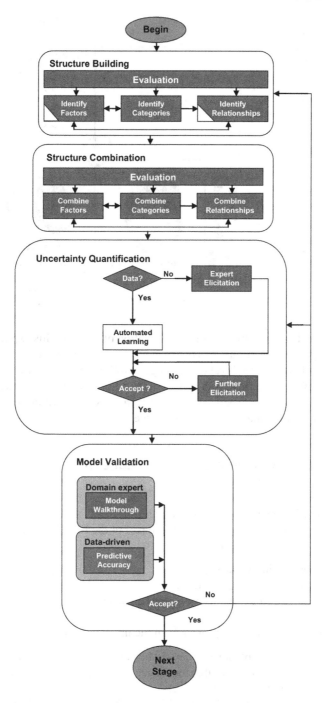

Fig. 6.4 Adapted expert-based knowledge engineering of Bayesian networks

Fig. 6.5 The three different stages of the theory of organisational knowledge creation that are used when building expert-based Bayesian network models. In this figure, I stands for Individual, G for group and O for Organisation

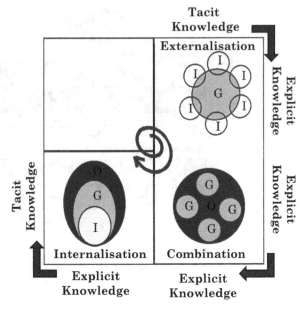

existing research and/or practice, or a combination of both. As this explicit knowledge is manipulated via discussions between the domain experts, it can be revisited, but it can also lead to some level of learning. Learning during the EKEBN process occurs more often when there is at least a slight difference in experiences between the participating domain experts. Whenever learning occurs within this context it represents an internationalisation of knowledge.

How Does the Internalisation of Knowledge Occur After Employing the EKEBN Process?

Once the Bayesian network model is built and validated, which generally happens after several iterations through the three-step part of the EKEBN process, it is used by different practitioners, such as:

- Project managers individually;
- Project managers working as a team;
- A project manager and other project team members (e.g., software developers); and
- Project managers and other project team members (e.g., software developers).

Such combination(s) of practitioners is not meant to be an exhaustive list. It simply reflects the combinations that took place when this book's author carried out the case studies detailed in Chaps. 7–12.

Whenever the organisation has practitioners who will use the effort estimation model despite not participating in the elicitation meetings used to build this model, they will need to be trained and to understand the combined mental model embedded in the Bayesian network model. This process also represents an internalisation of knowledge as the explicit knowledge incorporated in the Bayesian network model will need to be understood so it can be employed to obtain effort estimates for new projects.

In summary, the theory of organisational knowledge creation prescribes four different stages of knowledge creation; two of these are employed (and a third one can also be employed whenever needed) when building expert-based Bayesian network models (Fig. 6.4).

Note that a very important aspect when employing the EKEBN process is continuity and focus. In all the case studies detailed in Chaps. 7–12 we held weekly meetings, each lasting for 3 h. They were all held at the participating companies' premises; however, we made it clear to all of these companies that any model was only as good as their commitment and also their experience. Participants therefore were asked to refrain from leaving the meeting room to deal with any other work-related issues.

Detailed Structure Building

The focus of this step is to elicit the factors and the relationships that the domain expert(s) participating in the elicitation sessions identify as important for effort estimation. As previously mentioned, it is important to start from somewhere, i.e., using some explicit knowledge as basis for the discussions. At the time when this book's author was collaborating with companies and helping them building effort prediction models using Bayesian networks, the only comprehensive research providing a detailed set of factors that were identified as suitable effort predictors for Web projects was the Tukutuku project [1]. We will also use the same factors herein, however, nowadays there is also a systematic literature review on Web resource estimation (presented in Chap. 4) that suggests a greater number of factors identified as suitable effort predictors; therefore any companies following our process can either use the same Tukutuku factors we will use in this book, or use the entire set of factors detailed in Chap. 4. The most important aspect is that an initial and relevant set of factors be used, rather than to initiate the work using a "blank slate".

In essence, in order to identify the fundamental factors that the domain expert(s) take into account when preparing a project quote we recommend that the set of variables from the Tukutuku dataset [1] be used as a starting point (Table 6.1).

Therefore the following phases should be followed when identifying the fundamental factors:

Table 6.1 The Tukutuku variables

	Variable name	Description
Project Data	TypeProj	Type of project (new or enhancement)
	nLang	Number of different development languages used
	DocProc	If project followed defined and documented process
	ProImpr	If project team involved in a process improvement programme
	Metrics	If project team part of a software metrics programme
	DevTeam	Size of a project's development team
	TeamExp	Average team experience with the development language(s) employed
Web application	TotWP	Total number of Web pages (new and reused)
	NewWP	Total number of new Web pages
	TotImg	Total number of images (new and reused)
	NewImg	Total number of new images created
	Num_Fots	Number of features reused without any adaptation
	HFotsA	Number of reused high-effort features/functions adapted
	Hnew	Number of new high-effort features/functions
	TotHigh	Total number of high-effort features/functions
	Num_FotsA	Number of reused low-effort features adapted
	New	Number of new low-effort features/functions
	TotNHigh	Total number of low-effort features/functions

Phase 1

Sketch the Tukukutu variables on a whiteboard (or any other medium that enables the display of these variables), each one inside an oval shape, leaving space in between them (example given in Fig. 6.6). Any tool (whiteboard or other) that enables the easy drawing and redrawing of factors helps keep the elicitation sessions flowing. It also helps with visualising very clearly the set of factors that are being selected, which is important when making tacit knowledge explicit. Our previous experience eliciting Bayesian network models in other domains (e.g., ecology) suggested that it was best to start with a few factors (even if they were not to be reused by the domain expert(s)), rather than to start from a blank canvas.

Phase 2

Once the variable have been sketched out, each factor should be explained to the participants. Their meaning is bound by what they represent within the context of the Tukutuku project.

Fig. 6.6 The Tukutuku variables sketched out

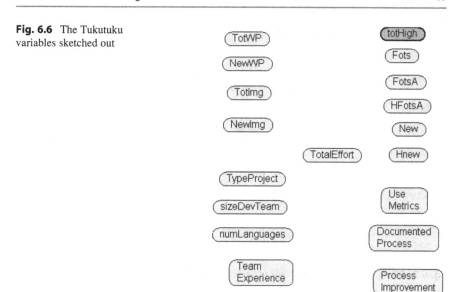

Phase 3

Once the Tukutuku variables had been sketched out and explained, the next phase is to remove all variables that the domain expert(s) do not consider relevant (example given in Fig. 6.7).

However, how to make this happen? Eliciting tacit knowledge is not an easy task as one needs to provide the means for the participants to "go deep" into the estimation process such that what they suggest does indeed reflect the factors they believe to be fundamental for effort estimation. In addition, there is also the issue when several domain experts are participating as there will be differences in their personalities (and also seniority in the company), and it is often the case that one wants to dominate the discussion and the decisions.

So let's take one step at a time.

In order to facilitate a fruitful elicitation exercise, ask the domain experts to think about their most recent effort estimation activity, and start from there. Doing it this way will take them back to a real and tangible scenario, which is important to ask suggest tangible and important factors, relationships and uncertainty quantifications (the latter is detailed later on). Once they are all there, focus on one of the Tukutuku factors at a time, and use a round robin style to get each of the domain experts to speak and voice their opinion on whether the factor under focus should be kept or not. Listen but also keep taking them back to their estimation scenario by consistently asking them to justify their opinion. In other words, it is very important to persistently ask "Why" so domain experts can clearly ground their suggestions. This also serves another purpose, which is to share their detailed tacit knowledge with the other domain experts who are also present.

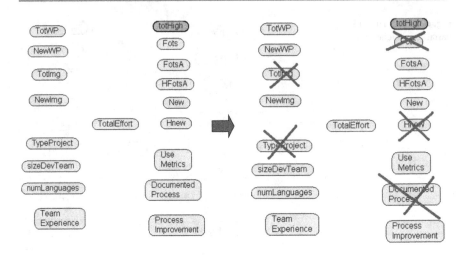

Fig. 6.7 Example showing the removal of several Tukutuku factors

Phase 4

Once Phase 3 is completed, the next phase comprises adding to the white board any additional variables (factors) suggested by the domain expert(s). The same approach of asking them to justify every decision should also be carried out here, so to get them grounded on the choices of factors being put forward. It is our experience that this step is occasionally slightly mixed with Phase 3, as domain experts may end up also suggesting new factors while they are considering whether to remove or not a Tukutuku factor, given that they are immersed thinking about their most recent effort estimation activity.

It is also very important to document the descriptions for each of the factors suggested. In addition, some of the initial meetings could also be recorded so to keep a memory of all the decisions taken. Such recording can also be used in case there are decision points that need further clarification. We have recorded all of our elicitation meetings with all the companies we collaborated with, and feel it is indeed a good practice to follow.

Let's suppose that the result of Phase 4 is the set of factors shown in Fig. 6.8. They have been grouped, whenever applicable, so to make it simpler to see how they relate to one another. The factors in light blue represent size-related effort predictors, and the ones in light purple represent cost drivers relating to people and the technology employed. Their description is given in Table 6.2.

These are the same factors previously used in the example model presented throughout Chap. 1, and their description and further details are presented again herein in order to help the reader follow the details presented herein.

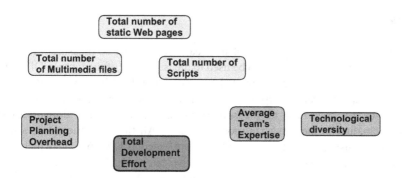

Fig. 6.8 Factors that were decided upon by the domain experts

Table 6.2 Description of the factors used in the example model

Factor	Description
Total number of static Web pages	This factor represents the estimated number of new Web pages that need to be created. These are not dynamically-generated pages, and include any types of page such as .htm, .html, .php
Total number of scripts	This factor represents the estimated sum of any types of scripts that are likely to be created for the Web application. Such scripts can be made using a client-side scripting technique (e.g., XML, Ajax techniques, Flash ActionScript), or server-side scripting languages (ASP, JSP, Perl, PHP, Python). It also includes files written using cascading style sheets (CSS)
Total number of multimedia files	This factor represents the total estimated number of any multimedia content, such as images and videos
Average team's expertise	This factor measures team expertise as the average number of years of experience that the development team has with Web development. The estimation within this context relates to tentative decision as to who will likely be allocated to the team that will develop the Web application for which total effort is being estimated
Technological diversity	This factor represents the estimated amount of diversity as far as the use of technology is concerned. It is measured using a surrogate measure, represented by the number of different technologies that are being employed in order to develop a Web application. Examples of technologies are MySQL, PHP, HTML, CSS, Python, ASP and JSP
Project planning overhead	This factor represents the degree of participation needed by the project manager in order to ensure the project is managed adequately and is ideally completed within time and on budget. This includes, but is not limited to status reports, communication, implementation plan (more for large projects) which includes the tasks to be done and their estimated completion dates, risk analysis, data analysis, planning (project execution plan)
Total development effort	This factor represents the total estimated effort to develop a Web application

Phase 5

This phase entails the definition of how each factor will be measured, i.e., how many and which categories are to be used to measure each of the factors. Of course, this decision will have an effect upon the number of probabilities to elicit during the uncertainty quantification step; however, such considerations should not constrain the choice of how best to measure each factor, as per the domain expert (s) participating in the Bayesian model building. All the categories must also be documented, using a template similar to that presented in Table 6.3. Note that in this example all the factors were measured using five categories; however, this is simply an example. Chapters 7–12 present real industrial Bayesian models and clearly show that the number of categories employed can vary quite widely within and across models.

Once all the categories are identified and documented, it is time to move on to the next phase, which is to elicit the cause and effect relationships between the factors.

Phase 6

As a starting point to this phase it is a good idea to use a simple example that explains in a simple way what is meant by a cause and effect relationship. When eliciting Bayesian network models with companies we have always used a simple medical example from Jensen [7] (Fig. 6.9). This simple model shows that there is a causal link between smoking and lung cancer (those who smoke are more likely to have lung cancer when compared to those who do not smoke) and the same in relation to lung cancer and coughing, and lung cancer and a positive x-ray for lung cancer. It also introduces one of the most important points to consider when identifying cause and effect relationships—the timeline of events. If smoking is to be a possible cause of lung cancer, it is important that the cause precedes the effect. This may sound obvious with regard to the example used; however, it is our view that the use of this simple example significantly helped the domain experts we have collaborated with understand the notion of cause and effect, and how this related to Web effort estimation and the Bayesian network models being elicited.

With regard to the factors shown in Fig. 6.8, their cause and effect relationships are shown in Fig. 6.10.

Figure 6.10 shows that all the factors affect "total development effort" directly; however, this is an issue when we think about the next step—uncertainty quantification.

Why would it be a problem? If we leave the model as is, the conditional probability table associated with the factor "total development effort" will contain $5 \times 5 \times 5 \times 5 \times 5 \times 5 \times 5$ probabilities. This means eliciting 78,125 probabilities manually, as there are not yet general solutions to generating probabilities semiautomatically, which can be readily incorporated to any Bayesian network tool.

Table 6.3 Description of the factors used in the example model and their corresponding categories

Factor	Description
Total number of static Web pages	This factor represents the estimated number of new Web pages that need to be created. These are not dynamically-generated pages, and include any types of page such as .htm, .html, .php. The total number of static Web pages is measured using five different categories (very large, large, medium, small and very small). This means that when a project manager is using the model, they have these five categories to choose from, and the choice will depend on the set of requirements they have gathered from the client to whom this application is to be developed. Within the context of the example model, these categories are detailed as follows: Very small number of Web pages → 0–5 Web pages Small number of Web pages → 6–15 Web pages Medium number of Web pages → 16–25 Web pages Large number of Web pages → 26–30 Web pages Very large number of Web pages → 31+ Web pages
Total number of scripts	This factor represents the estimated sum of any types of scripts that are likely to be created for the Web application. Such scripts can be made using a client-side scripting technique (e.g., XML, Ajax techniques, Flash ActionScript), or server-side scripting languages (ASP, JSP, Perl, PHP, Python). It also includes files written using cascading style sheets. The total number of scripts is measured using five different categories (very large, large, medium, small and very small). Within the context of the example model, these categories are detailed as follows: Very small number of scripts/css files → 0–7 scripts/css files Small number of scripts/css files → 8–20 scripts/css files Medium number of scripts/css files → 21–25 scripts/css files Large number of scripts/css files → 26–35 scripts/css files Very large number of scripts/css files → 36+ scripts/css files
Total number of multimedia files	This factor represents the total estimated number of any multimedia content, such as images and videos. The total number of multimedia files is measured using five different categories (very large, large, medium, small and very small). Within the context of the example model, these categories are detailed as follows: Very small number of multimedia content → 0–3 multimedia content Small number of multimedia content → 4–8 multimedia content Medium number of multimedia content → 9–20 multimedia content Large number of multimedia content → 21–30 multimedia content Very large number of multimedia content → 31+ multimedia content
Average team's expertise	This factor measures team expertise as the average number of years of experience that the development team has with Web development. The estimation within this context relates to tentative decision as to who will likely be allocated to the team that will develop the Web application for which total effort is being estimated Five different categories (very high, high, average, low and very low) are used to measure average team's expertise. Within the context of the example model, these categories are detailed as follows: Very low team's expertise → 1 year of experience Low team's expertise → 2–3 years of experience Average team's expertise → 4–8 years of experience High team's expertise → 9–12 years of experience Very high team's expertise → 13+ years of experience

(continued)

Table 6.3 (continued)

Factor	Description
Technological diversity	This factor represents the estimated amount of diversity as far as the use of technology is concerned. It is measured using a surrogate measure, represented by the number of different technologies that are being employed in order to develop a Web application. Examples of technologies are MySQL, PHP, HTML, CSS, Python, ASP and JSP. The five categories employed to measure Technological diversity are Within the context of the example model, these categories are detailed as follows: Very low technological diversity → 1 type of technology is being used in the Web application Low technological diversity → 2 different types of technology are being used in the Web application Average technological diversity → 3–4 different types of technology are being used in the Web application High technological diversity → 5–7 different types of technology are being used in the Web application Very high technological diversity → 8+ different types of technology are being used in the Web application
Project planning overhead	This factor represents the degree of participation needed by the project manager in order to ensure the project is managed adequately and is ideally completed within time and on budget. This includes, but is not limited to, status reports, communication, implementation plan (more for large projects) which includes the tasks to be done and their estimated completion dates, risk analysis, data analysis, planning (project execution plan) The project planning overhead is measured in our example model using five different categories (very high, high, average, low and very low), which are detailed as follows: Very low project overhead → 5 % of estimated effort Low project overhead → 15 % of estimated effort Average project overhead → 20 % of estimated effort High project overhead → 30 % of estimated effort Very high project overhead → 40 % of estimated effort
Total development effort	This factor represents the total estimated effort to develop a Web application. The three factors that have a direct effect upon total effort are: combined cost factors' effort, project planning overhead and combined size's effort. The total development effort is also measured in our example model using five different categories (very high, high, average, low and very low), which are detailed as follows: Very low effort → 1–126 person-hours Low effort → 126+ to 320 person-hours Average effort → 320+ to 670 person-hours High effort → 670+ to 1,400 person-hours Very high effort → 1,400+ person-hours

Is there another solution to this issue? Yes, there is. We need to create interme-diate factors which goal is simply to reduce the number of relationships targeting a particular factor. The factor that is the focus herein is "total development effort". We name these intermediate factors Optimisation factors. Figure 6.11 shows the

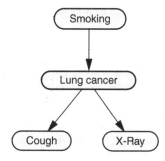

Fig. 6.9 A simple example illustrating cause and effect relationships

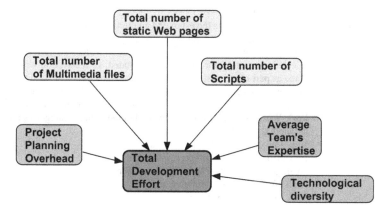

Fig. 6.10 Cause and effect relationships for the example model

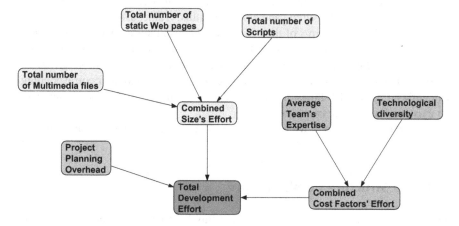

Fig. 6.11 Final structure for the example model

Table 6.4 Description of the two optimisation factors used in the example model and their corresponding categories

Factor	Description
Combined size's effort	This factor represents the estimated amount of effort (person-hours) needed to create Web pages, scripts/css files and multimedia files. Note that the effort will change depending on which categories are selected for each factor. Such selection will take place as part of a decision making scenario, and examples will be given later on in this chapter. The combined size's effort is measured using five different categories (very high, high, average, low and very low). Within the context of the example model, these categories are detailed as follows: Very low effort → 1–40 person-hours Low effort → 40+ to 80 person-hours Average effort → 80+ to 160 person-hours High effort → 160+ to 320 person-hours Very high effort → 320+ person-hours
Combined cost factors' effort	This factor represents the estimated amount of effort (person-hours) when taking into account technological diversity and average team's expertise. Note that the effort will change depending on which categories are selected for each factor. As previously stated, such selection will take place as part of a decision making scenario, and examples will be given later on in this chapter. The combined cost factors' effort is measured using five different categories (very high, high, average, low and very low). Within the context of the example model, these categories are detailed as follows: Very low effort → 1–80 person-hours Low effort → 80+ to 200 person-hours Average effort → 200+ to 400 person-hours High effort → 400+ to 800 person-hours Very high effort → 800+ person-hours

final structure of our example model, which contains two optimisation factors—combined size's effort and combined cost factors' effort.

Note that whenever new factors are included in the model, we also need to define how they are to be measured, and document this. Table 6.4 describes the two optimization factors in terms of their meaning and also their associated categories.

Once all the categories have been defined to each of the factors, it is time to use any available Bayesian network tool that a company has access to, so an initial model can be created containing factors, their relationships and their corresponding categories. Figure 6.12 shows the same model shown in Fig. 6.11, now created with an existing Bayesian network tool.

We are now ready to move on to the next step—uncertainty quantification.

Detailed Structure Combination

This step relates to the merging of different Bayesian network structures in readiness for the next step. It only applies to companies or organisations that have different groups building their separate Bayesian network model structures and

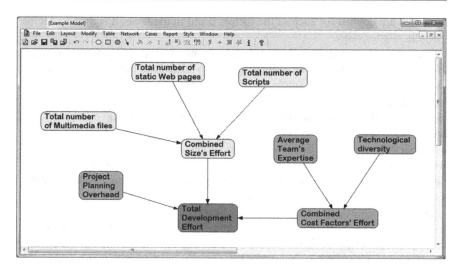

Fig. 6.12 Final structure for the example model

who are willing to combine these structures into a single, company/organisation-wide model structure, in order to obtain a common model.

Each of the model structures will contain factors, the categories used to measure each of them and how they relate to one another. These factors and categories should also have been documented by the teams working on each of the separate model structures. Therefore, the very first task to be carried out is to do the semantic mapping across all factors and categories. This is important in order to identify which factors are common across different model structures.

We suggest the use of a table where rows represent factors and columns represent model structures. Model structures can be labelled as MS1... MS*n*, where *n* corresponds to the total number of model structures being combined. Whenever the same factor is represented in different model structures by different names it is important to agree upon which name to use. This is the name to be included in a row. We would also suggest that a document is kept recording all the decisions that are made while combining the different model structures. For example, all the alternative names that were used for a factor, how each of those was measured in terms of categories and which one was selected to be used in the combined model structure (and corresponding categories).

Once all factors are entered as rows and mapped to all the model structures they originally came from, it is time to work through each of the factors and decide how the factor will be measured, based on the choice of categories that were put forward by the different groups. We do not have a suggestion as to the ideal number of categories to use in order to measure a factor. This will really depend on what is considered most important for the company or organisation. Some want more precision, so they use at least seven categories to measure a factor; others are happy with measuring factors using three to five categories only. Once all categories are defined for each of the factors listed in the table, it is time to decide upon the relationships.

We suggest first targeting the relationships and corresponding factors directly affecting the factor that is to be estimated. Within the context of this book this factor is total effort. Using as an example the model structure shown in Fig. 6.12, the three factors directly affecting effort are those originating from factors "combined size's effort", "combined cost factors' effort", and "project planning overhead" and directly targeting the factor "total development effort". This is to be done with one model structure at a time, where one should only change to another model structure once the previous one has been dealt with completely. So this means that this task needs to be carried out sequentially, working through one model structure at a time.

Once all the factors and relationships directly pointing to effort have been defined, it is time to move to those factors that directly affect the factors that directly affect effort. In terms of the model structure shown in Fig. 6.12, they would correspond to factors "total number of static Web pages", "total number of scripts", and "total number of multimedia files". The idea is to move each time further and further away from the target factor being estimated.

During this process there may be situations where two different relationships contradict themselves (e.g., using the model structure in Fig. 6.12 as an example, in one model structure (MS1) factor "total number of multimedia files" has only a relationship to factor "total number of static Web pages", and in another model structure (MS10) factor "total number of multimedia files" has only a relationship to factor "combined size's effort"). Whenever this happens, the team working on the combination of model structures will have to decide which relationship to keep, or whether all the relationships should be represented in the combined model structure. It is also very important to remove any possible cycles in the combined model (e.g., factor A has a relationship to factor B, and factor B has a relationship back to factor A) as Bayesian network models cannot present cycles due to the causal nature of their relationships.

Detailed Uncertainty Quantification

This step relates to the quantification of the uncertainty inherent to the complex domain being modelled, which is done using probabilistic reasoning. All quantifications are provided using the conditional probability tables that are associated with every factor part of a Bayesian network model; however, the mechanism used to reason about the probabilities for the "parent" factors (factors that are not directly affected by any others) differs slightly from the mechanism applied to the other factors (those that are affected directly by at least another factor). Let's see how this works by explaining how to quantify the probabilities for two different factors part of our example model—total number of static Web pages and combined cost factors' effort.

Total number of static Web pages—this factor is measured using five categories (very large, large, medium, small and very small), so when eliciting the probabilities from the participating domain experts, we need to ask them questions such as the following:

- If you consider all the past projects you managed as a pie, what slice of this pie (what percentage of projects) presented a very large number of static Web pages (31+ Web pages)? Why?
- If you consider all the past projects you managed as a pie, what slice of this pie (what percentage of projects) presented a large number of static Web pages (26–30 Web pages)? Why?
- If you consider all the past projects you managed as a pie, what slice of this pie (what percentage of projects) presented a medium number of static Web pages (16–25 Web pages)? Why?
- If you consider all the past projects you managed as a pie, what slice of this pie (what percentage of projects) presented a small number of static Web pages (6–15 Web pages)? Why?
- If you consider all the past projects you managed as a pie, what slice of this pie (what percentage of projects) presented a very small number of static Web pages (1–5 Web pages)? Why?

It is clear that these questions are all asking for the frequency of occurrence of projects with a particular number of static Web pages. This will always be the case when quantifying uncertainty relating to parent factors. Let's assume that the answers to these questions are respectively 10, 30, 40, 10 and 10. The conditional probability table relating to the factor total number of static Web pages will therefore look like the one shown in Fig. 6.13.

Note that after every question asked we also added a "Why?" question. This is done in order to motivate domain experts to explain their reasoning. Our experience shows that whenever further explanations (more detailed explanations) need to be provided, domain experts may change their answers as a result of having to explain in detail the reasoning behind the answer that was provided.

Now, let's look at the other type of factor using as example the factor "combined cost factors' effort".

Combined cost factors' effort—this factor is measured using five categories (very large, large, medium, small and very small), and is directly affected by another two factors—average team's expertise and technological diversity. This means that in order to reason about any quantification relating to "combined cost factors' effort", we need to also take into account the other two factors that are affecting it. Therefore, the conditional probability table associated with "combined cost factors' effort" will look similar to the one shown in Fig. 6.14. Note that the conditional probability table shown in Fig. 6.14 already presents example probabilities. In a real situation this table would be empty, and the domain experts would fill it out with the uncertainty quantifications.

Very Large	Large	Medium	Small	Very Small
10	30	40	10	10

Fig. 6.13 Conditional probability table for factor "total number of static Web pages"

Technological diversity	Average Team's Expertise	Very High	High	Average	Low	Very Low
Very High	Very High	0	0	0	25	75
Very High	High	0	0	5	25	70
Very High	Average	70	25	5	0	0
Very High	Low	85	15	0	0	0
Very High	Very Low	90	10	0	0	0
High	Very High	0	0	0	30	70
High	High	0	0	15	25	60
High	Average	60	25	15	0	0
High	Low	70	30	0	0	0
High	Very Low	80	20	0	0	0
Average	Very High	0	0	0	40	60
Average	High	0	0	35	25	40
Average	Average	40	25	35	0	0
Average	Low	60	40	0	0	0
Average	Very Low	70	30	0	0	0
Low	Very High	0	0	0	10	90
Low	High	0	0	5	25	70
Low	Average	0	0	40	35	25
Low	Low	0	0	70	30	0
Low	Very Low	0	5	75	20	0
Very Low	Very High	0	0	0	10	90
Very Low	High	0	0	0	30	70
Very Low	Average	0	0	40	40	20
Very Low	Low	0	0	50	50	0
Very Low	Very Low	0	0	60	40	0

Fig. 6.14 Conditional probability table for factor "combined cost factors' effort"

This time, when eliciting the probabilities from the participating domain experts, we would need to ask them questions such as the following:

- Please **only** consider past projects for which the "technological diversity" was very high and the "average team's expertise" was very high. Assume this to be your only set of projects to think about. If this set of projects was a pie, what percentage of this pie would correspond to a very high "combined cost factors' effort"? Why?
- Please **only** consider past projects for which the "technological diversity" was very high and the "average team's expertise" was very high. Assume this to be your only set of projects to think about. If this set of projects was a pie, what percentage of this pie would correspond to a high "combined cost factors' effort"? Why?
- Please **only** consider past projects for which the "technological diversity" was very high and the "average team's expertise" was very high. Assume this to be

your only set of projects to think about. If this set of projects was a pie, what percentage of this pie would correspond to a average "combined cost factors' effort"? Why?

- Please **only** consider past projects for which the "technological diversity" was very high and the "average team's expertise" was very high. Assume this to be your only set of projects to think about. If this set of projects was a pie, what percentage of this pie would correspond to a low "combined cost factors' effort"? Why?

- Please **only** consider past projects for which the "technological diversity" was very high and the "average team's expertise" was very high. Assume this to be your only set of projects to think about. If this set of projects was a pie, what percentage of this pie would correspond to a very low "combined cost factors' effort"? Why?

Each of these questions has to take the domain expert(s) to a *decision scenario* corresponding to the specific combination of categories from the two factors that are affecting "combined cost factors' effort".

It is also very important here to ask domain experts to explain *why* they believe a given combination of categories (from the factors that are affecting "combined cost factors' effort") will have a particular probability of causing an effect over "combined cost factors' effort". As they explain, they are working on their tacit level, and by listening to their own explanations they may revisit the probabilities that were initially suggested. We have witnessed this numerous times during our experience building several Bayesian network models in collaboration with companies. We also tend to use a pie as a metaphor for reasoning as it has been previously suggested in the literature as a good technique to employ [9].

> **Important**
> What happens when there are certain combinations of categories that do not make any sense in practice? For example, very low "technological diversity" and very low "average team's expertise" representing a decision scenario that never happens for a given company. Our proposed solution would be to add an extra category to factor "combined cost factors' effort", where this category would only be used to deal with "impossible" scenarios. Such category could be called "Never", "Nil", or whichever name the domain expert(s) feel best represents an impossible scenario.

Once all the quantifications for all the conditional probability tables have been elicited, the model is ready to be validated. Figure 6.15 shows our example model with factors also displaying the bars corresponding to all the probability quantifications that were done.

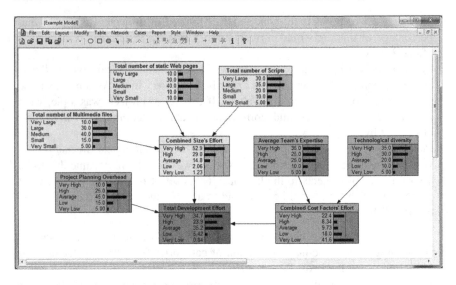

Fig. 6.15 Example model ready for validation

Detailed Model Validation

This step represents the validation of the Bayesian model that was built via the two steps previously detailed. It generally comprises two activities—model walk-through and predictive accuracy—where both aim to check how well calibrated the Bayesian model is.

Model walk-through is a calibration that is done subjectively, where domain experts use several hypothetical scenarios to check, for each of these scenarios, whether the effort category that the model shows with the highest probability matches the category they would have selected if estimating effort subjectively. We see this activity as a way to enable domain experts to get a feel for the model, and therefore we do not recommend that companies rely solely on this activity in order to decide whether the model is ready for use.

Predictive accuracy entails the use of past data from completed projects, for which total development effort is known, in order to check the model's calibration. We recommend that most of these completed projects—called the validation set—be representative of the typical projects developed by the company. Table 6.5 provides an example, which will be used herein, showing past data for three projects. This data will be used next to validate our example model. Note that we are showing the hypothetical data for each of the projects based on the same categories used in the example model just to keep this discussion focused. In reality, there could already be some existing data from past projects stored as numbers. In that case each number would need to be "translated" into one of the categories used

Table 6.5 An example of three past projects that hypothetically represent "typical" projects

Past completed project	Total number of static Web pages	Total number of scripts	Total number of multimedia files	Average team's expertise	Technological diversity	Project planning overhead	Total development effort
1	Very large	Medium	Medium	Average	Very high	High	Very high
2	Very small	Very large	Medium	High	High	Low	Average
3	Very small	Very large	Medium	High	High	Very low	Average

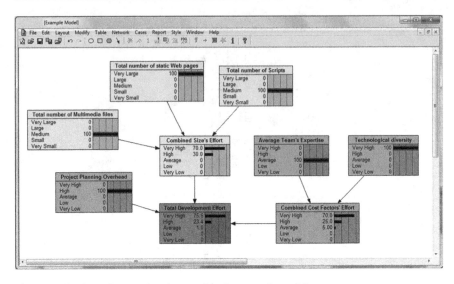

Fig. 6.16 Evidence from project 1 entered in the example model

by the same corresponding factor in the Bayesian model, prior to carrying out the predictive accuracy activity.

For each project, evidence needs to be entered in the example model. Figure 6.16 shows the model after the known data from project 1 was entered. Note that "entering data" represents clicking on the category that corresponds to the project data available. By doing so, we are also acknowledging that within the context of that scenario we are "certain" about the category that was selected. This is represented in the model by a change to 100 (100 %) associated with the selected category, and a change to 0 for all the other categories that are part of that same factor.

Figure 6.16 shows that "very high" is the category for factor "total development effort" that presents the highest probability (75.6 %). Does this result match the actual data for this same project, which is displayed in Table 6.5?

Table 6.5 shows that for project 1, the total development effort was "very high", which matches the category suggested by the example model. This means the model is in line with the data from that past project, which translates as being calibrated, given that particular decision scenario.

Let's now enter evidence from project 2 in the example model (Fig. 6.17).

Figure 6.17 shows that "average" is the category for factor "total development effort" that presents the highest probability (66 %). Does this result match the actual data for this same project, which is displayed in Table 6.5?

Table 6.5 shows that for project 2, the total development effort was "average", which matches the category suggested by the example model. Again, this means that the model is in line with the data from that past project, which also translates as being calibrated given that particular decision scenario.

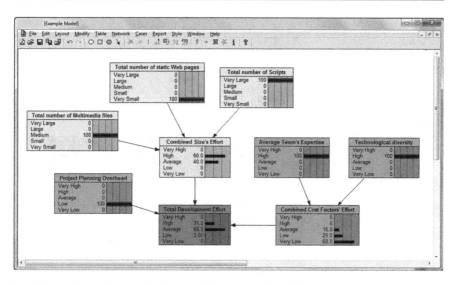

Fig. 6.17 Evidence from project 2 entered in the example model

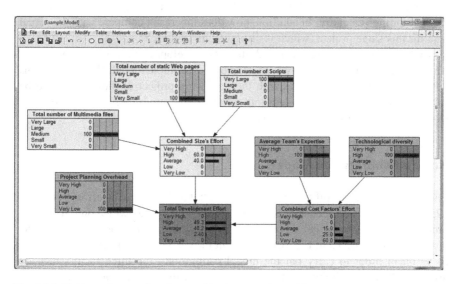

Fig. 6.18 Evidence from project 3 entered in the example model

Finally, let's enter evidence for the last project detailed in Table 6.5—project 3 (Fig. 6.18).

Figure 6.18 shows that the category "high" for factor "total development effort" presents the highest probability (49.3 %). However, the model also shows that there is still quite a large uncertainty in relation to this choice, given that the category

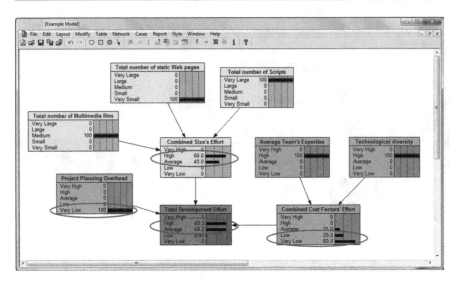

Fig. 6.19 How to identify which decision scenarios to change in the conditional probability table

"average" presents a very similar probability (48.2 %). This suggests that there is no clear "winning" category. In addition, how does this result compare with the actual data for this same project, which is displayed in Table 6.5?

Table 6.5 shows that for project 3, the total development effort was "average", which differs from the category suggested by the example model. What this means is that the model is not in line with the data from project 3; thus it needs to be calibrated for that particular decision scenario, and any others that the domain experts feel that should also be revisited.

How to Calibrate the Model?

The conditional probability table for factor "total development effort" contains 125 different decision scenarios, so how can we know exactly which ones to look in order to calibrate the model according to the specific scenario shown in Fig. 6.18?

The red oval shapes in Fig. 6.19 highlight the categories and factors we need to focus on in order to carry out the model calibration—the three factors that are affecting "total development effort" directly, and their corresponding categories presenting the highest (and optionally, also the second-highest) probabilities. These will correspond to the decision scenarios in the conditional probability table for factor "total development effort" shown in Fig. 6.20, highlighted with red rectangles.

Figure 6.20 shows that the first two decision scenarios highlighted in red present the highest probabilities for the effort category "high" (90 and 75); the last two decision scenarios highlighted in red present the highest probabilities to the effort category "average", so already matching the actual effort data for project 3. This

[Total_Effort Table (in net New_effort_example)]
File Edit Table Window Help

Node: Total_Effort

Chance   % Probability

Apply Okay
Reset Close

Project Planning Overhead	Combined Size's Effort	Combined Cost Factors' Effort	Very High	High	Average	Low	Very Low
Very Low	Very High	Very High	100	0	0	0	0
Very Low	Very High	High	100	0	0	0	0
Very Low	Very High	Average	95	5	0	0	0
Very Low	Very High	Low	90	5	5	0	0
Very Low	Very High	Very Low	80	15	5	0	0
Very Low	High	Very High	100	0	0	0	0
Very Low	High	High	10	90	0	0	0
Very Low	High	Average	0	95	5	0	0
Very Low	High	Low	0	90	10	0	0
Very Low	High	Very Low	0	75	25	0	0
Very Low	Average	Very High	95	5	0	0	0
Very Low	Average	High	0	95	5	0	0
Very Low	Average	Average	0	5	95	0	0
Very Low	Average	Low	0	0	100	0	0
Very Low	Average	Very Low	0	0	90	10	0
Very Low	Low	Very High	90	5	5	0	0
Very Low	Low	High	0	90	10	0	0
Very Low	Low	Average	0	0	100	0	0
Very Low	Low	Low	0	0	5	95	0
Very Low	Low	Very Low	0	0	0	100	0
Very Low	Very Low	Very High	80	15	5	0	0

Fig. 6.20 Decision scenarios in the conditional probability table for factor "total development effort"

[Total_Effort Table (in net New_effort_example)]
File Edit Table Window Help

Node: Total_Effort

Chance   % Probability

Apply Okay
Reset Close

Project Planning Overhead	Combined Size's Effort	Combined Cost Factors' Effort	Very High	High	Average	Low	Very Low
Very Low	Very High	Very High	100	0	0	0	0
Very Low	Very High	High	100	0	0	0	0
Very Low	Very High	Average	95	5	0	0	0
Very Low	Very High	Low	90	5	5	0	0
Very Low	Very High	Very Low	80	15	5	0	0
Very Low	High	Very High	100	0	0	0	0
Very Low	High	High	10	90	0	0	0
Very Low	High	Average	0	95	5	0	0
Very Low	High	Low	0	35	65	0	0
Very Low	High	Very Low	0	25	75	0	0
Very Low	Average	Very High	95	5	0	0	0
Very Low	Average	High	0	95	5	0	0
Very Low	Average	Average	0	5	95	0	0
Very Low	Average	Low	0	0	100	0	0
Very Low	Average	Very Low	0	0	90	10	0
Very Low	Low	Very High	90	5	5	0	0
Very Low	Low	High	0	90	10	0	0
Very Low	Low	Average	0	0	100	0	0
Very Low	Low	Low	0	0	5	95	0
Very Low	Low	Very Low	0	0	0	100	0
Very Low	Very Low	Very High	80	15	5	0	0

Fig. 6.21 Calibrated decision scenarios in the conditional probability table for factor "total development effort"

means that only the first two decision scenarios need to be calibrated, leading to the conditional probability table shown in Fig. 6.21.

Now, when we run the same decision scenario again, the category that shows the highest probability for "total development effort" matches the actual effort for project 3 (Fig. 6.22).

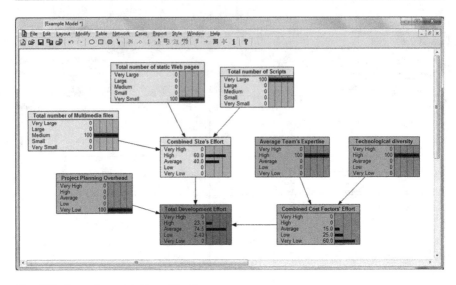

Fig. 6.22 Example model after calibration

Whenever a calibration occurs, it is recommended that the model be rechecked again for all the projects that were already used for validation, just to ensure that the calibration did not have a negative knock-on effect.

When to stop validating a model? The decision when to stop validating a model is really dependent on the amount of past project data available, and also on how confident domain experts are that the choice and range of projects used to validate a model are enough. We have dealt with companies where domain experts were happy with using data from only 8 past projects, to companies where domain experts used data from 22 past projects.

In our opinion the more data is used for validation the better; however, this is also clearly dependent on how much data is available. If a model is large and complex (e.g., more than 30 factors) we believe it is very important to use a reasonable number of past data to start with (data from at least 20 or so projects), and to carry on validating the model as new projects are completed. This is more pressing for larger models; however, this practice can equally apply to smaller projects too.

Models such as the example model are not meant to be stagnant once validated. Current practices, types of projects and applications can change over time and therefore it is important to revisit the model so to keep it up to date with those changes.

Conclusions

This chapter has detailed the steps that are part of the expert-based knowledge engineering of Bayesian networks process, and has also related these steps with the theory of organisational knowledge creation.

The three main steps within the expert-based knowledge engineering of Bayesian networks process (EKEBN) are the structure building, uncertainty quantification, and model validation. This process iterates over these steps until a complete Bayesian network model is built and validated. Structure building represents the identification of factors, relationships and how each factor is going to be measured, i.e., its categories. Uncertainty quantification relates to populating all the conditional probability tables associated with each of the factors previously identified. Finally, model validation represents the validation of the model using scenarios and also past data.

References

1. Mendes E, Mosley N, Counsell S (2005) Investigating Web size metrics for early Web cost estimation. J Syst Softw 77(2):157–172
2. Mendes E, Mosley N (2008) Bayesian network models for Web effort prediction: a comparative study. IEEE Trans Softw Eng 34(6):723–737
3. Woodberry O, Nicholson A, Korb K, Pollino C (2004) Parameterising Bayesian networks. In: Proceedings of the Australian conference on artificial intelligence, pp 1101–1107
4. Studer R, Benjamins VR, Fensel D (1998) Knowledge engineering: principles and methods. Data Knowl Eng 25:161–197
5. Mendes E, Pollino C, Mosley N (2009) Building an expert-based Web effort estimation model using Bayesian networks. In: Proceedings of the EASE conference, pp 1–10
6. Baker S, Mendes E (2010) Aggregating expert-driven causal maps for Web effort estimation. In: Proceedings of the advances in software engineering (ASEA) conference: communications in computer and information science, vol 117, pp 264–282. doi: 10.1007/978-3-642-17578-7_27
7. Jensen FV (1996) An introduction to Bayesian networks. UCL Press, London
8. Nonaka I, Toyama R (2003) The knowledge-creating theory revisited: Knowledge creation as a synthesizing process. Knowl Manag Res Pract 1:2–10
9. Druzdzel MJ, van der Gaag LC (2000) Building probabilistic networks: where do the numbers come from? IEEE Trans Knowl Data Eng 12(4):481–486

Effort and Risk Prediction for Healthcare Software Projects Delivered on the Web

7

Introduction

This chapter revisits the expert-based knowledge engineering of Bayesian networks (EKEBN) process that was detailed in Chap. 6 (Fig. 7.1), describing the tasks carried out for each of the three main steps that form part of that process. Before starting the elicitation of the healthcare effort and risk-prediction BN model, the domain experts (DEs) participating were presented with an overview of Bayesian network models, and examples of "what-if" scenarios using a made-up BN. This, we believe, facilitated the entire process as the use of an example, and the brief explanation of each of the steps in the EKEBN process, provided a concrete understanding of what to expect. We also made it clear that the knowledge engineer was a facilitator of the process, and that the healthcare company's commitment was paramount for the success of the process.

The entire process took 324 person-hours to complete, with seven people participating in twelve 3-h slots, and two people participating in other twelve 3-h slots.

The DEs who took part in this case study were project managers of a well-established healthcare company in Auckland (New Zealand). This company represents one of the several branches worldwide that are part of a larger healthcare organization, which headquarters in Japan. The company had ~70 employees. The project managers had each worked in healthcare software development for more than 10 years. In addition, this company developed a wide range of healthcare software applications, using different types of technology.

Detailed Structure Building and Uncertainty Quantification

In order to identify the fundamental factors that the DEs took into account when preparing a project quote, we used the set of variables from the Tukutuku dataset [1] as a starting point (Table 7.1). We first sketched them out on a whiteboard, each one inside an oval shape, and then explained what each one meant within the context of

E. Mendes, *Practitioner's Knowledge Representation*, DOI 10.1007/978-3-642-54157-5_7, 107
© Springer-Verlag Berlin Heidelberg 2014

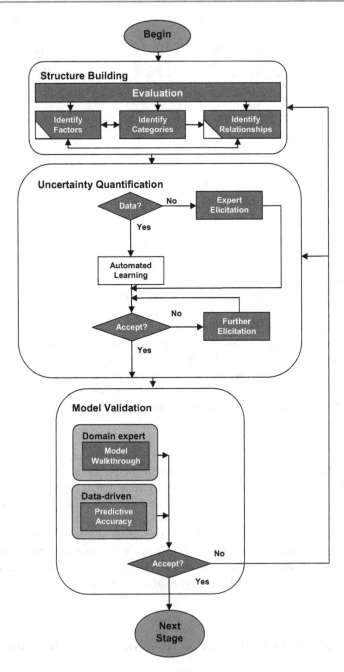

Fig. 7.1 Expert-based knowledge engineering of Bayesian networks process

Table 7.1 Tukutuku variables

	Variable name	Description
Project data	*TypeProj*	Type of project (new or enhancement)
	nLang	Number of different development languages used
	DocProc	If project followed defined and documented process
	ProImpr	If project team involved in a process improvement programme
	Metrics	If project team part of a software metrics programme
	DevTeam	Size of a project's development team
	TeamExp	Average team experience with the development language(s) employed
Web application	*TotWP*	Total number of Web pages (new and reused)
	NewWP	Total number of new Web pages
	TotImg	Total number of images (new and reused)
	NewImg	Total number of new images created
	Num_Fots	Number of features reused without any adaptation
	HFotsA	Number of reused high-effort features/functions adapted
	Hnew	Number of new high-effort features/functions
	TotHigh	Total number of high-effort features/functions
	Num_FotsA	Number of reused low-effort features adapted
	New	Number of new low-effort features/functions
	TotNHigh	Total number of low-effort features/functions

the Tukutuku project. Our previous experience eliciting BNs in other domains (e.g., ecology) suggested that it was best to start with a few factors (even if they were not to be reused by the DE), rather than to use a "blank canvas" as a starting point.

Within the context of the Tukutuku project, a new high-effort feature/function requires at least 15 h to be developed by one experienced developer, and a high-effort adapted feature/function requires at least 4 h to be adapted by one experienced developer. These values are based on collected data.

Once the Tukutuku variables had been sketched out and explained, the next step was to remove all variables that were not relevant for the DEs, followed by adding to the whiteboard any additional variables (factors) suggested by them. We also documented descriptions for each of the factors suggested. Next, we identified the states that each factor would take. All states were discrete. Whenever a factor represented a measure of effort (e.g., total effort), we also documented the effort range corresponding to each state, to avoid any future ambiguity. For example, "very low" total effort corresponded to 4+ to 10 person-hours, etc. Once all states were identified and documented, it was time to elicit the cause and effect relationships. As a starting point to this task we used a simple medical example from [2] (Fig. 7.2).

This example clearly introduces one of the most important points to consider when identifying cause and effect relationships—the timeline of events. If smoking is to be a cause of lung cancer, it is important that the cause precedes the effect. This may sound obvious with regard to the example used; however, it is our view that the

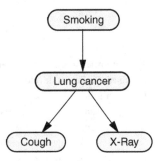

Fig. 7.2 A small BN causal structure

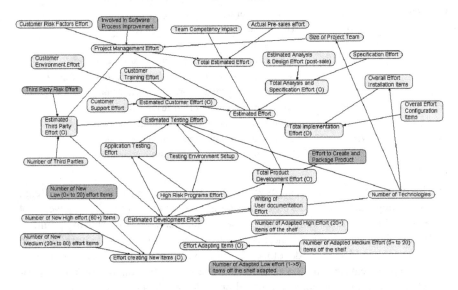

Fig. 7.3 Expert-based healthcare software causal structure

use of this simple example significantly helped the DEs understand the notion of cause and effect, and how this related to software effort and risk estimation and the BN being elicited.

Once the cause and effect relationships were identified, the healthcare software effort and risk BN's causal structure was as follows (Fig. 7.3). Note that Fig. 7.3 is not a BN based directly on Table 7.1. During this process several factors were each reached by a large number of relationships; therefore the model's initially proposed structure needed to be simplified in order to reduce the number of probabilities to be elicited. New factors were suggested by the KE (factor names ending in "(O)", see Fig. 7.3), and validated by the DEs. This is the final structure presented in Fig. 7.3. It contains 38 factors and 37 corresponding relationships that were identified by the DEs as fundamental for software effort and risk prediction.

The description of each of the factors used in the healthcare software effort and risk estimation BN model is given in Table 7.2.

Table 7.2 Description of all the factors elicited from the DEs

Factor	Categories	Description, observation
Actual presales effort	Low (0+ to 10 person-hours) Medium (10+ to 20 person-hours) High (20+ person-hours)	Contract signing (optional), requirements elicitation (prepared before preparation of quote) + quote preparation, user requirements specification (optional), programming specification (aka technical spec, functional spec)
Number of technologies	Small (1 technology) Medium (2–3 technologies) Large (4 and above)	Examples of internal technologies: Cobol, Web (ASP, .NET, C#), Windows, Lotus Notes, Oracle, SQL
Application testing effort	None Low (0+ to 10 person-hours) Medium (10+ to 30 person-hours) High (30+ to 150 person-hours) Very high (150+ person-hours)	Testing throughout the project, but only inside the company
Testing environment setup	Low (0+ to 1 person-hours) Medium (1+ to 4 person-hours) High (4+ person-hours)	Number of person-hours to set up the testing environment
High risk programs effort	None Low (0+ to 5 person-hours) Medium (5+ to 10 person-hours) High (10+ to 20 person-hours) Very high (20+ person-hours)	Programs used by only a few customers, and difficult to test; programs that are historically difficult to manage or change (e.g., nondocumented features, COBOL legacy)
Estimated third-party effort	None Low (0+ to 10 person-hours) Medium (10+ to 30 person-hours) High (30+ to 60 person-hours) Very high (60+ person-hours)	Estimated effort to third party-related issues (number and risk factor)
Effort adapting items	None Very low (0+ to 10 person-hours) Low (10+ to 20 person-hours) Medium (20+ to	Number of person-hours adapting items

(continued)

Table 7.2 (continued)

Factor	Categories	Description, observation
	40 person-hours) High (40+ to 80 person-hours) Very high (80+ person-hours)	
Effort creating new items	None Low (0+ to 40 person-hours) Medium (40+ to 80 person-hours) High (80+ to 150 person-hours) Very high (150+ to 1,000 person-hours) Extremely high (1,000+ person-hours)	Number of person-hours creating new items
Effort to create and package product	None Low (0+ to 1 person-hour) Medium (1+ to 4 person-hours) High (4+ person-hours)	Effort to create and package a product (includes paperwork, burning a CD, printing and binding the manuals, issuing the product (send the CD to the customer, or uploading into a FTP site)); also includes maintaining internal source code repository, and patches
Writing of user documentation effort	None Low (0+ to 10 person-hours) Medium (10+ to 50 person-hours) High (50+ to 200 person-hours) Very high (200+ person-hours)	Estimate of the number of hours writing the user documentation (aka product documentation, user manual)
Estimated testing effort	Low (0+ to 10 person-hours) Medium (10+ to 30 person-hours) High (30+ to 150 person-hours) Very high (150+ to 450 person-hours) Extremely high (450+ person-hours)	Total estimated testing effort from environment set up and application testing
Estimated development effort	None Very Low (0+ to 20 person-hours) Low (20+ to 80 person-hours) Medium (80+ to 150 person-hours)	Total estimated development effort from the items

(continued)

Table 7.2 (continued)

Factor	Categories	Description, observation
	High (150+ to 450 person-hours) Very high (450+ to 1,000 person-hours) Extremely high (1,000+ person-hours)	
Total product development effort	None Very low (0+ to 20 person-hours) Low (20+ to 80 person-hours) Medium (80+ to 150 person-hours) High (150+ to 450 person-hours) Very high (450+ to 2,500 person-hours) Exceptionally high (2,500+ person-hours)	
Customer environment effort	Low (0+ to 1 person-hour) Medium (1+ to 5 person-hours) High (5+ person-hours)	Time zone, system access; these are tangible points
Customer risk factors effort (generally represented as an effort %)	None Low (0+ to 5 person-hours) Medium (5+ to 10 person-hours) High (10+ person-hours)	Personality, capabilities, expectations, involvement, track record, language barrier, language difficulties, size customer representation/team
Customer support effort	None Low (0+ to 8 person-hours) Medium (8+ to 40 person-hours) High (40+ person-hours)	Pre- and post-go live support
Customer training effort	None Low (0+ to 8 person-hours) Medium (8+ to 40 person-hours) High (40+ person-hours)	Amount of training (includes preparation)
Estimated customer effort	None Low (0+ to 20 person-hours) Medium (20+ to 85 person-hours) High (85+ person-hours)	Estimated effort for customer-related items (environment, support, training)

(continued)

Table 7.2 (continued)

Factor	Categories	Description, observation
Involved in software process improvement (SPI)	Yes No	Part of the project management
Number of adapted high effort (20+) items off-the-shelf	None Small (1 item) Medium (2 items) High (3+ items)	Number of hours that represent high and low effort need to be defined (excludes testing). One adaptation can incur several changes. High effort here means the use of 20+ person-hours to adapt a single item
Number of adapted medium effort (5+ to 20) items off-the-shelf	None Small (1 item) Medium (2–4 items) High (5+ items)	Number of hours that represent high and low effort need to be defined (excludes testing). One adaptation can incur several changes. Medium effort here means the use of 5+ to 20 person-hours to adapt a single item
Number of adapted low effort (1→5) items off-the-shelf	None Small (1–3 items) Medium (4–6 items) High (7+ items)	(Excludes testing) One adaptation can incur several changes. Low effort here means the use of up to 5 person-hours to adapt a single item
Number of new high effort (80+) items	None Small (1 item) Medium (2–4 items) High (5+ items)	(Excludes testing) High effort here means the use of 80+ person-hours to develop a single item
Number of new low effort items	None Small (1 item) Medium (2–4 items) High (5+ items)	(Excludes testing) Low effort here means the use of up to 20 person-hours to develop a single item
Number of new medium effort items	None Small (1 item) Medium (2–4 items) High (5+ items)	(Excludes testing) Medium effort here means the use of 20+ to 80 person-hours to develop a single item
Overall effort configuration items	None Very low (0+ to 1 person-hours) Low (1+ to 5 person-hours) Medium (5+ to 15 person-hours) High (15+ to 40 person-hours) Very high (40+ person-hours)	Effort to configure an installed system for use as per customer requirements
Overall effort installation items	None Low (0+ to 5 person-hours) Medium (5+ to 15 person-hours) High (15+ person-hours)	Items are interpreted as an area, program or module. Items have hour figures next to them. (Either it's only development, or pure training, CD sent to client for them to install)

(continued)

Table 7.2 (continued)

Factor	Categories	Description, observation
Total implementation effort	None Very low (0+ to 2 person-hours) Low (2+ to 5 person-hours) Medium (5+ to 20 person-hours) High (20+ to 80 person-hours) Very high (80+ person-hours)	
Project management effort	None Low (15 % of estimated effort) Medium (20–30 % of estimated effort) High (30+ % of estimated effort)	Project management overhead, including status reports; communication; implementation plan (more for large projects) which includes the tasks to be done and their estimated completion dates; risk analysis; data analysis; planning (project execution plan)
Size of project team	Small (2–5 people) Medium (6–8 people) Large (9+ people)	Only the team internally to the company
Estimated analysis and design effort (post-sales)	None Low (0+ to 5 person-hours) Medium (5+ to 20 person-hours) High (20+ to 70 person-hours) Very high (70+ person-hours)	Requirements elicitation, user requirements specification, programming specification (aka technical spec, functional spec)
Specification effort	None Low (0+ to 3 person-hours) Medium (3+ to 10 person-hours) High (10+ person-hours)	Set-up plan, cut-over plan (steps required to move changes into production), customer test specification
Total analysis and specification effort	None Low (0+ to 8 person-hours) Medium (8+ to 30 person-hours) High (30+ to 80 person-hours) Very high (80+ person-hours)	

(continued)

Table 7.2 (continued)

Factor	Categories	Description, observation
Team competency impact	Very low (0 % of the team have low competency) Low (0 % + to 25 % of the team have low competency) Medium (25 % + to 40 % of the team have low competency) High (40 % + to 70 % of the team have low competency) Very high (70 % + to 100 % of the team have low competency)	Definition to be considered when rating: – years of experience with the domain (e.g., hematology), – years of experience with programming language, technical skill – knowledge of the product, (Y/N) – training (not charged to the customer), – technology (development technology and target technology, e.g., virtual environment), – non-SNZ team members (Y/N) – English as a second language (Y/N) – software development lifecycle role – proven past performance – customer/market knowledge (e.g., when writing specifications) – personality (e.g., attention to detail, easily distracted, note: this is often only known after the project) – experience in development and implementation of beta products
Third-party risk effort	None Low (0+ to 5 person-hours) Medium (5+ to 10 person-hours) High (10+ person-hours)	Not company's customers (for example, emailing third party, phone calls, finalising specs, reading their documentation, communication plan, messages)
Number of third parties	None Small (1 third party) Medium (2–3 third parties) High (4 or more third parties)	Number of external systems (sw, hw) or organisations (people, third parties company has no control over)
Estimated effort	None Very low (0+ to 15 person-hours) Low (15+ to 40 person-hours) Medium (40+ to 150 person-hours) High (150+ to 1,000 person-hours) Very high (1,000+ to 3,000 person-hours) Exceptionally high (3,000+ person-hours)	Estimated effort to develop a project, excluding project management

(continued)

Table 7.2 (continued)

Factor	Categories	Description, observation
Total estimated effort	None Very low (0+ to 15 person-hours) Low (15+ to 40 person-hours) Medium (40+ to 150 person-hours) High (150+ to 1,500 person-hours) Very high (1,500+ to 4,000 person-hours) Exceptionally high (4,000+ person-hours)	Total estimated effort to develop a project, including project management

At this point the DEs seemed happy with the BN's causal structure and the work on eliciting the probabilities was initiated. All probabilities were created from scratch, and the probabilities elicitation took 72 h (one DE and one KE). The complete BN, including its probabilities, is shown in Fig. 7.4. Figure 7.4 shows the BN using belief bars rather than labelled factors, so readers can see the probabilities that were elicited.

Detailed Model Validation

Both model walk-through and predictive accuracy were used to validate the healthcare software effort and risk prediction BN model, where the former was the first type of validation to be employed. The DE used ten different scenarios to check whether the factor "total estimated effort" would provide the highest probability to the effort state that corresponded to the DE's own suggestions. All scenarios were run successfully; however, it was also necessary to use data from past projects, for which total effort was known, in order to check the model's calibration. A validation set containing data on 22 projects was used. The DE selected a range of projects presenting different sizes and levels of complexity, where all 22 projects were representative of the types and sizes of projects developed by the healthcare company.

For each project, evidence was entered in the BN model (an example is given in Fig. 7.5, where evidence is characterised by dark grey factors with probabilities equal to 100 % (1. . .)), and the effort range corresponding to the highest probability provided for "total estimated effort" was compared to that project's actual effort. For example, in Fig. 7.6, this would correspond to "total estimated effort" = high. The company had also defined the range of effort values associated with each of the categories used to measure "total estimated effort". In the case of the company described herein, high effort corresponded to 150–1,500 person-hours. Whenever actual effort did not fall within the effort range associated with the category with the highest probability, there

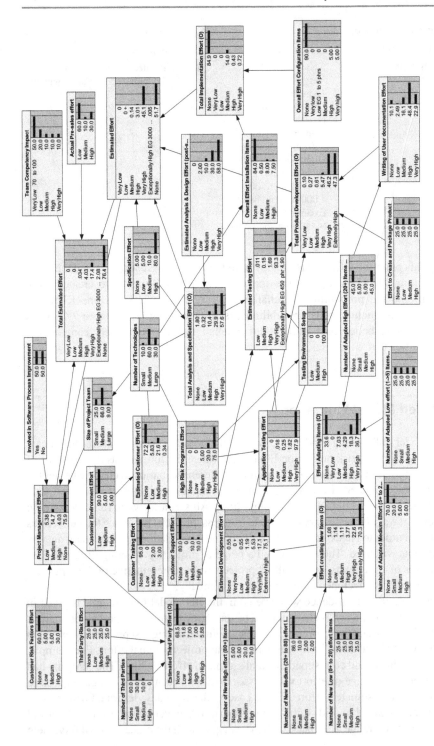

Fig. 7.4 Expert-based healthcare software BN

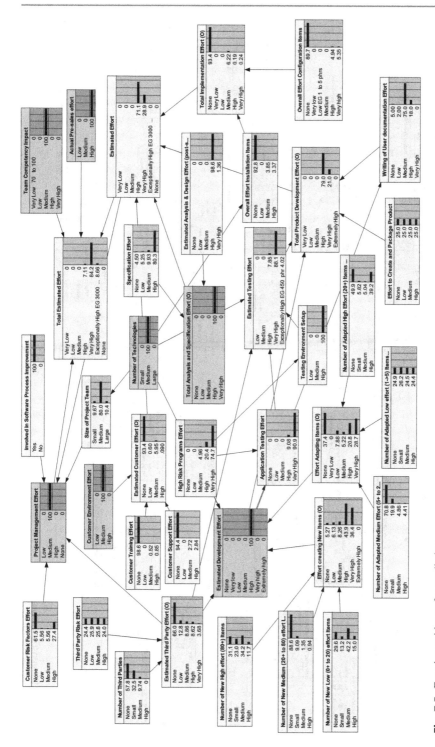

Fig. 7.5 Entering evidence for prediction

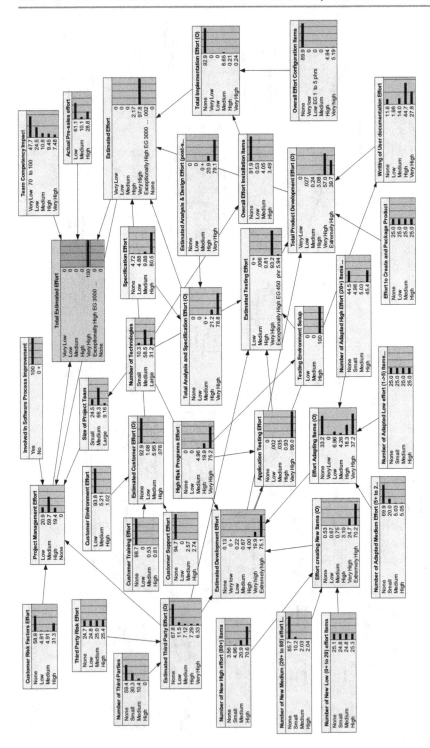

Fig. 7.6 Diagnostic reasoning

was a mismatch; this meant that some probabilities needed to be adjusted. Within the context of this work, hardly any calibration was needed.

Whenever probabilities were adjusted, we re-entered the evidence for each of the projects in the validation set that had already been used in the validation step to ensure that the calibration already carried out had not affected. This was done to ensure that each calibration would always be an improvement upon the previous one. Within the scope of the model presented herein, of the 22 projects used for validation, only one required the model to be recalibrated. This means that for all the 21 projects remaining, the BN model presented the highest probability to the effort range that contained the actual effort for the project being used for validation. Once all 22 projects were used to validate the model the DE assumed that the validation step was complete.

In terms of the use of this BN model, it can also be employed for diagnostic reasoning, and to run numerous "what-if" scenarios. Figure 7.6 shows an example of a model being used for diagnostic reasoning, where the evidence was entered for total estimated effort, and used to assess the highest probabilities for each of the other factors.

The BN model was completed in February 2010, and has been successfully used to estimate effort and risks for new healthcare software projects developed by the company. In addition, the DE who participated in the causal structure and probabilities' elicitation completely changed her approach to estimating effort as follows: she presented the BN model to all of her development team, and asked that from that point onwards every estimate for any task would need to be based on the factors that had been elicited. This means that the entire team started to use the factors that were elicited, as well as the BN model, as basis for their effort and risk-estimation sessions. In addition, the DE presented the model at a meeting with other branches, so to detail how the Auckland branch was estimating effort and risk for their healthcare projects. The other branches were so impressed, in particular the one from the US, that they increased the number of healthcare software projects outsourced to the NZ branch, as they recognized the benefits of using a model that represented factors and uncertainties. Overall, such change in approach provided extremely beneficial to the company.

We believe that the successful development of this healthcare software effort and risk estimation BN model was greatly influenced by the commitment of the company, and also by the DEs' exceptional experience estimating effort.

Conclusions

This chapter has presented a case study where a Bayesian model for effort and risk estimation of healthcare projects was built using solely knowledge of six domain experts from a well-established healthcare company in Auckland, New Zealand. This model was developed using the expert-based knowledge engineering for Bayesian networks process (Fig. 7.4). Each session with the DEs lasted for no longer than 3 h. The final BN model was calibrated using data on 22 past projects. These projects represented typical projects developed by the company, and believed by the experts to provide enough data for model calibration.

Since the model's adoption, it has been successfully used to provide effort quotes for the new projects managed by the company.

The entire process used to build and validate the BN model took 324 person-hours, used as follows: 252 person-hours for the first 12 weeks (6 DEs + 1 KE); 72 h for the last 12 weeks (1 DE + 1 KE).

The elicitation process enables experts to think deeply about their effort and risk estimation process and the factors taken into account during that process, which in itself is already advantageous to a company. This has been pointed out to us not only by the DEs whose model is presented herein, but also by other companies with which we worked on model elicitations.

References

1. Mendes E, Mosley N, Counsell S (2005) Investigating web size metrics for early web cost estimation. J Syst Softw 77(2):157–172
2. Jensen FV (1996) An introduction to Bayesian networks. UCL Press, London

Effort Prediction for Multimedia Projects Delivered on the Web

Introduction

This chapter revisits the expert-based knowledge engineering of Bayesian networks (EKEBN) process that was detailed in Chap. 6 (Fig. 8.1), describing the tasks carried out for each of the three main steps that form part of that process. Before starting the elicitation of the Multimedia-based BN model, the domain experts (DEs) participating were presented with an overview of Bayesian network models, and examples of "what-if" scenarios using a made-up BN. This, we believe, facilitated the entire process as the use of an example, and the brief explanation of each of the steps in the EKEBN process, provided a concrete understanding of what to expect. We also made it clear that the knowledge engineer was a facilitator of the process, and that the company's commitment was paramount for the success of the process.

The entire process took 66 person-hours to be completed, with two people participating in eleven 3-h slots.

The domain expert who took part in this case study was the project manager of a well-established Web company in Auckland (New Zealand). At the time the model was built, the company had ~20 employees. The project manager had worked in multimedia and Web development for more than 15 years. In addition, this company also developed a wide range of kiosk software applications, using different types of technology.

Detailed Structure Building and Uncertainty Quantification

In order to identify the fundamental factors that the DEs took into account when preparing a project quote, we used the set of variables from the Tukutuku dataset [1] as a starting point (Table 8.1). We first sketched them out on a board, each one inside an oval shape, and then explained what each one meant within the context of the Tukutuku project. Our previous experience eliciting BNs in other domains (e.g.,

E. Mendes, *Practitioner's Knowledge Representation*, DOI 10.1007/978-3-642-54157-5_8,   123
© Springer-Verlag Berlin Heidelberg 2014

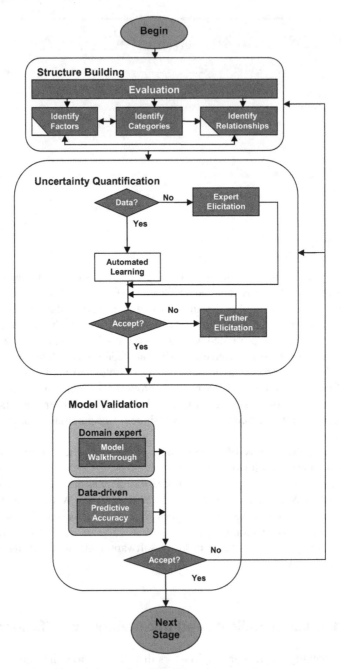

Fig. 8.1 Expert-based knowledge engineering of Bayesian networks process

Table 8.1 Tukutuku variables

	Variable name	Description
Project data	*TypeProj*	Type of project (new or enhancement)
	nLang	Number of different development languages used
	DocProc	If project followed defined and documented process
	ProImpr	If project team involved in a process improvement programme
	Metrics	If project team part of a software metrics programme
	DevTeam	Size of a project's development team
	TeamExp	Average team experience with the development language(s) employed
Web application	*TotWP*	Total number of Web pages (new and reused)
	NewWP	Total number of new Web page
	TotImg	Total number of images (new and reused)
	NewImg	Total number of new images created
	Num_Fots	Number of features reused without any adaptation
	HFotsA	Number of reused high-effort features/functions adapted
	Hnew	Number of new high-effort features/functions
	TotHigh	Total number of high-effort features/functions
	Num_FotsA	Number of reused low-effort features adapted
	New	Number of new low-effort features/functions
	TotNHigh	Total number of low-effort features/functions

ecology) suggested that it was best to start with a few factors (even if they were not to be reused by the DE), rather than to use a "blank canvas" as a starting point.

Within the context of the Tukutuku project, a new high-effort feature/function requires at least 15 h to be developed by one experienced developer, and a high-effort adapted feature/function requires at least 4 h to be adapted by one experienced developer. These values are based on collected data.

Once the Tukutuku variables had been sketched out and explained, the next step was to remove all variables that were not relevant for the DEs, followed by adding to the whiteboard any additional variables (factors) suggested by them. We also documented descriptions for each of the factors suggested. Next, we identified the states that each factor would take. All states were discrete. Whenever a factor represented a measure of effort (e.g., total effort), we also documented the effort range corresponding to each state, to avoid any future ambiguity. For example, "low level 1" total effort corresponded to 18–40 person-hours, etc. Once all states were identified and documented, it was time to elicit the cause and effect relationships. As a starting point to this task we used a simple medical example from [2] (Fig. 8.2).

This example clearly introduces one of the most important points to consider when identifying cause and effect relationships—the timeline of events. If smoking is to be a cause of lung cancer, it is important that the cause precedes the effect. This may sound obvious with regard to the example used; however, it is our view that the

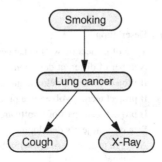

Fig. 8.2 A small BN illustrating cause and effect relationships

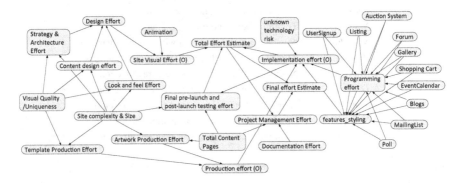

Fig. 8.3 Expert-based Web effort causal structure

use of this simple example significantly helped the DEs understand the notion of cause and effect, and how this related to software effort and risk estimation and the BN being elicited.

Once the cause and effect relationships were identified, the Web effort BN's causal structure was as shown in Fig. 8.3. Note that Fig. 8.3 is not a BN based directly on Table 8.1. During this process the factors "final effort estimate" and "total effort estimate" were each reached by a large number of relationships; therefore this structure needed to be simplified in order to reduce the number of probabilities to be elicited. New factors were suggested by the KE (factor names ending in "(O)", see Fig. 8.3), and validated by the DE. This is the final structure presented in Fig. 8.3. A total of 31 factors and 55 corresponding relationships were identified as influential to effort estimation.

Each of the factors used in the Web effort estimation Multimedia-based BN model is presented in Table 8.2. Whenever a description is missing, this is because the factor's name is already self-explanatory.

At this point the DE seemed happy with the BN's causal structure, and the work on eliciting the probabilities was initiated. Most of the probabilities were created from scratch; however, some were also obtained from existing data on past finished Web projects. The probabilities elicitation subprocess took 36 h (one DE and one

Table 8.2 Factors identified by the domain experts

Factor	Categories	Description, observation
Template production effort	Very low (8+ to 16 person-hours) Low (16+ to 24 person-hours) Medium (24+ to 40 person-hours) Medium high (40+ to 60 person-hours) High (60+ to 90 person-hours) Very high (90+ to 120 person-hours) Super high (120+ person-hours)	Effort making (producing) all the different templates for all the different pages
Artwork production effort	Very low (0+ to 8 person-hours) Low (8+ to 16 person-hours) Medium (16+ to 24 person-hours) Medium high (24+ to 40 person-hours) High (40+ to 60 person-hours) Very high (60+ to 80 person-hours) Super high (80+ person-hours)	Effort making (producing) all the artwork for all the different pages
Team experience	25 % 50 % 75 % 100 %	Percentage of team members with optimal experience
Tight deadline	Yes, no	
Unknown technology risk	Yes, no	
External hosting	Yes, no	
Overall risk	Low, medium, high	
Content design effort	Very low (0+ to 8 person-hours) Low (8+ to 24 person-hours) Medium (24+ to 40 person-hours) Medium-high (40+ to 60 person-hours) High (60+ person-hours)	
Visual quality/uniqueness	Template standard, Template high,	Uniqueness of the application's visual quality

(continued)

Table 8.2 (continued)

Factor	Categories	Description, observation
	custom-medium, custom-high	
Site complexity and size	Small (1 audience group/topic) Medium (2 audience groups/topics) Medium-large (3 audience groups/topics) Large (4 audience groups/topics) Very large (5+ audience groups/topics)	Number of different types of sections (content areas/types of functionality; different audience groups). The greater the number of different types of users, the greater the care with providing functionality and content areas that are suitable to each type of user. It involves identifying how each part of the site will suit its audience, but making it all cohesive
Strategy and architecture effort	Very low (0+ to 8 person-hours) Low (8+ to 16 person-hours) Medium (16+ to 24 person-hours) High (24+ to 40 person-hours) Very high (40+ to 60 person-hours) Super high (60+ person-hours)	How to make sure a user doesn't get lost; how do you make sure you give your audience what they want, that they will find what they need. The architecture represents the navigation (providing landscape points to enable people to navigate without getting lost). The strategy represents deciding on the best mechanisms to enable users to find what they need quickly
Look and feel effort	Very low (0+ to 8 person-hours) Low (8+ to 24 person-hours) Medium-low (24+ to 40 person-hours) Medium (40+ to 60 person-hours) High (60+ person-hours)	Web branding design and art direction
Design effort	Very low (0+ to 16 person-hours) Low (16+ to 40 person-hours) Medium (40+ to 80 person-hours) Medium-high (80+ to 124 person-hours) High (124+ to 160 person-hours) Very high (160+ person-hours)	

(continued)

Table 8.2 (continued)

Factor	Categories	Description, observation
Site visual effort	Very low (0+ to 24 person-hours) Low (24+ to 56 person-hours) Medium (56+ to 104 person-hours) Medium-high (104+ to 164 person-hours) High (164+ to 220 person-hours) Very high (220+ person-hours)	
Total content pages (assumes that the client has provided the content and images)	1 to10 pages 11–20 pages 20–35 pages 35–50 pages 51–75 pages 76–100 pages 101–125 pages 126–250+ pages	
Programming effort	Very, very Low (1.5–4 person-hours) Very low (4+ to 12 person-hours) Low (12+ to 20 person-hours) Medium-low (20+ to 40 person-hours) Medium (40+ to 80 person-hours) Medium-high (80+ to 120 person-hours) High (120+ to 200 person-hours) Very high (200+ to 400 person-hours) Very, very high (400+ to 600 person-hours)	Represents the effort used to implement or adapt the features that will be part of a Web application (e.g., forum, gallery, shopping cart)
Production effort	Super, very low (8+ to 24 person-hours) Low (24+ to 40 person-hours) Medium (40+ to 64 person-hours) Medium-high (64+ to 100 person-	

(continued)

Table 8.2 (continued)

Factor	Categories	Description, observation
	hours) High (100+ to 150 person-hours) Very high (150+ to 200 person-hours) Super high (200+ person-hours)	
Documentation effort	Little (0+ to 10 person-hours) Medium (10+ to 20 person-hours) High (20–60 person-hours) New sort (60+ person-hours)	Applies to when some documentation needs to be created for the client
Animation	None Very low (0+ to 8 person-hours) Low (8+ to 16 person-hours) Medium (16+ to 24 person-hours) Medium-high (24+ to 40 person-hours) High (40+ to 60 person-hours) Very high (60+ to 100 person-hours) Super high (100+ person-hours)	
Final prelaunch and post-launch testing effort	Low (0+ to 12 person-hours) Medium low (12+ to 20 person-hours) Medium (20+ to 80 person-hours) High (80+ to 140 person-hours) Extremely high (140+ person-hours)	
Client approvals and communications	Yes, no	Client difficulty
Features styling (additional effort styling the features)	Very low (0+ to 4 person-hours) Low (4+ to 12 person-hours) Medium (12+ to 30 person-hours) High (30+ to	It represents the effort needed to adapt, for example, style sheets to take all the widgets of a particular feature (e.g., shopping cart) into account. A simplistic example would be if site is to be pink, then styling represents to ensure that all the features added to the

(continued)

Table 8.2 (continued)

Factor	Categories	Description, observation
	64 person-hours) Very high (64+ to 120 person-hours) Very, very high (120+ to 160 person-hours)	site comply with this requirement— being pink
Implementation effort	Very low (11+ to 48 person-hours) Low (48+ to 80 person-hours) Medium-low (80+ to 130 person-hours) Medium (130+ to 224 person-hours) Medium-high (224+ to 340 person-hours) High (340+ to 550 person-hours) Very high (550+ to 1,000 person-hours) Very, very high (1,000+ to 1,400 person-hours) Super high (1,400+ person-hours)	Represents the effort to adapt features. If a given feature needs to be developed from scratch they will estimate it outside this model
Project management effort	Very low (0+ to 10 person-hours) Low (10+ to 15 person-hours) Medium low (15+ to 30 person-hours) Medium (30+ to 40 person-hours) Medium high (40+ to 50 person-hours) High (50+ to 70 person-hours) Very high (70+ person-hours)	
Final effort estimate	Low level 1 (18+ to 40 person-hours), Low level 2 (40+ to 80 person-hours), Medium level 1 (80+ to 140 person-hours), Medium level 2 (140+ to	

(continued)

Table 8.2 (continued)

Factor	Categories	Description, observation
	200 person-hours) Medium level 3 (200+ to 300 person-hours) Medium-high level 1 (300+ to 500 person-hours) Medium-high level 2 (500+ to 800 person-hours), High level 1 (800+ to 1,000 person-hours), High level 2 (1,000+ to 1,300 person-hours), High level 3 (1,300+ to 1,500 person-hours), High level 4 (1,500+ to 1,700 person-hours), High level 5 (1,700+ person-hours)	
Total effort estimate	Low level 1 (18+ to 40 person-hours) Low level 2 (40+ to 80 person-hours) Medium level 1 (80+ to 140 person-hours) Medium level 2 (140+ to 200 person-hours) Medium level 3 (200+ to 300 person-hours) Medium-high level 1 (300+ to 500 person-hours) Medium-high level 2 (500+ to 800 person-hours) High level 1 (800+ to 1,000 person-hours) High level 2 (1,000+ to 1,300 person-hours) High level 3 (1,300+	

(continued)

Table 8.2 (continued)

Factor	Categories	Description, observation
	to 1,500 person-hours) High level 4 (1,500+ to 1,700 person-hours) High level 5 (1,700+ person-hours)	
Final effort estimate with risk	Low level 1 (18+ to 43 person-hours) Low level 2 (43+ to 87 person-hours) Medium level 1 (87+ to 150 person-hours) Medium level 2 (150+ to 215 person-hours) Medium level 3 (215+ to 320 person-hours) Medium-high level 1 (320+ to 530 person-hours) Medium-high level 2 (530+ to 840 person-hours) High level 1 (840+ to 1,040 person-hours) High level 2 (1,040+ to 1,350 person-hours) High level 3 (1,350+ to 1,570 person-hours) High level 4 (1,570+ to 2,000 person-hours) High level 5 (2,000+ person-hours)	
UserSignup	Yes, no	Users can sign up to the website and create their own accounts
Forum	Yes, no	
Auction system	Yes, no	
Listing (classified ads, etc., property listings)	None One Two Three	

(continued)

Table 8.2 (continued)

Factor	Categories	Description, observation
Gallery	None	
	One	
	Two	
	Three	
Shopping cart	Yes, no	
Event calendar	Yes, no	Displays a calendar control on the website; events can be added to it
Blogs (same as news)	None	
	One	
	Two	
	Three	
Poll	None	
	One	
	Two	
Mailing List	Yes, no	

KE). The complete BN, including its probabilities, is shown in Fig. 8.4. Figure 8.4 shows the BN using belief bars rather than labelled factors, so readers can see the probabilities that were elicited.

Detailed Model Validation

Both model walk-through and predictive accuracy were used to validate the Web effort estimation BN model, where the former was the first type of validation to be employed. The DE used ten different scenarios to check whether the factor "final effort estimate" would provide the highest probability to the effort state that corresponded to the DE's own suggestions. All scenarios were run successfully; however, it was also necessary to use data from past projects, for which total effort was known, in order to check the model's calibration. A validation set containing data on 22 projects was used. The DE selected a range of projects presenting different sizes and levels of complexity, where all 22 projects were representative of the types and sizes of projects developed by the Web company.

For each project, evidence was entered in the BN model (an example is given in Fig. 8.5, where evidence is characterised by dark grey factors with probabilities equal to 100 % (1...)), and the effort range corresponding to the highest probability provided for "final effort estimate" was compared to that project's actual effort. For example, in Fig. 8.5, this would correspond to "final effort estimate" = medium level 3. The company had also defined the range of effort values associated with each of the categories used to measure "final effort estimate". In the case of the company described herein, medium level 3 corresponded to 200+ to 300 person hours. Whenever actual effort did not fall within the effort range associated with the category with the highest probability, there was a mismatch; this meant that some

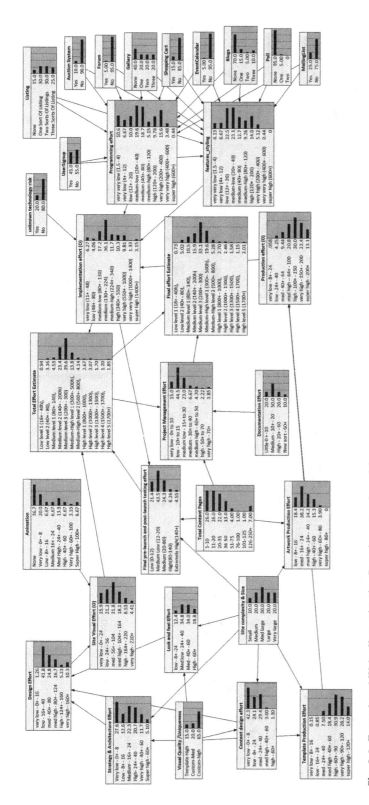

Fig. 8.4 Expert-based Web effort estimation Bayesian network model

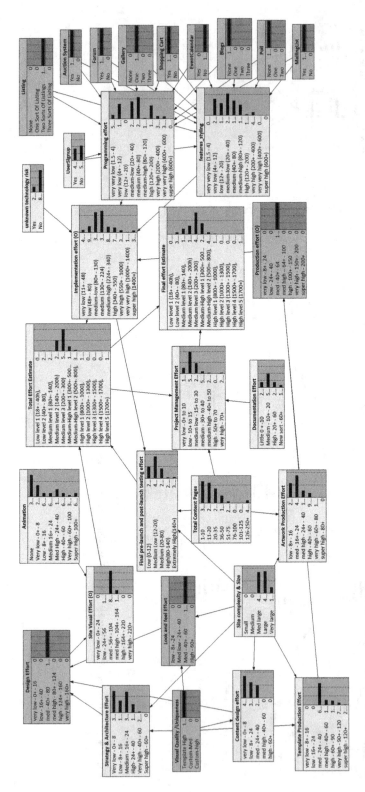

Fig. 8.5 Entering evidence for prediction

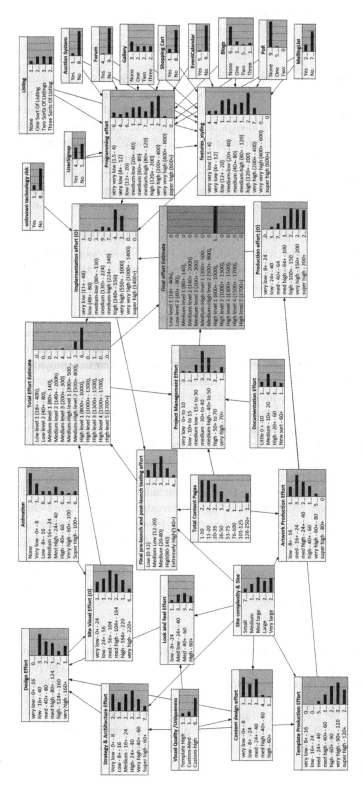

Fig. 8.6 Diagnostic reasoning

probabilities needed to be adjusted. Within the context of this work, hardly any calibration was needed.

Whenever probabilities were adjusted, we re-entered the evidence for each of the projects in the validation set that had already been used in the validation step to ensure that the calibration already carried out had not been affected. This was done to ensure that each calibration would always be an improvement upon the previous one. Within the scope of the model presented herein, of the 22 projects used for validation, only one required the model to be recalibrated. This means that for all the 21 projects remaining, the BN model presented the highest probability to the effort range that contained the actual effort for the project being used for validation. Once all 22 projects were used to validate the model the DE assumed that the validation step was complete.

In terms of the use of this BN model, it can also be employed for diagnostic reasoning, and to run numerous "what-if" scenarios. Figure 8.6 shows an example of a model being used for diagnostic reasoning, where the evidence was entered for "final effort estimate", and used to assess the highest probabilities for each of the other factors.

The BN model was completed in March 2010, and has been successfully used to estimate effort for new Web projects developed by the company. Prior to using the model, the company that is the focus of this chapter did not even know what set of factors they considered fundamental when estimating effort for their new projects; therefore the elicitation of factors and their causal relationships alone was already considered very helpful to them. In addition, they found it extremely useful to be able to run numerous "what-if" scenarios to help with their decision making, and in addition, to also be able to obtain a range of possible effort values and their associated uncertainty. These were very useful in order to negotiate project costs with clients, given that the effort estimates were based on much more solid knowledge than simply their tacit knowledge.

The factors that were identified by the DE did not include any of the factors used when applying a function points methodology to measuring size, because this company did not measure size using function points. However, the methodology that has been presented herein would equally apply to companies that employ function points.

We believe that the successful development of this Web effort estimation BN model was greatly influenced by the commitment of the company, and also by the DE's experience estimating effort.

Conclusions

This chapter has presented a case study where a Bayesian model for Web effort estimation was built using knowledge from a domain expert and also data on past finished Web projects developed by the company. This model was developed using the knowledge engineering for Bayesian networks process (see Fig. 8.1). Each session with the DE lasted for no longer than 3 h. The final BN model was calibrated using data on 22 past projects. These projects represented typical Web projects developed by the company, and were believed by the expert to provide enough data for model calibration. Since the model's adoption, it has been successfully used to provide effort quotes for the new Web projects managed by the company.

We have developed other BN models that were validated using data ranging from 8 to 12 past projects only. According to our experience building BNs for effort estimation, the most important aspect to obtain a sound model relates to the domain experts' knowledge of the effort estimation domain. Experienced experts will build models that require very little validation.

The entire process used to build and validate the BN model took 66 person-hours, used as follows: 24 person-hours for the first 4 weeks (1 DE + 1 KE); 42 person-hours for the last 7 weeks (1 DE + 1 KE).

The elicitation process enables experts to think deeply about their effort estimation process and the factors taken into account during that process, which in itself is already advantageous to a company. This has been pointed out to us not only by the domain expert whose model is presented herein, but also by other companies with which we worked on model elicitations.

References

1. Mendes E, Mosley N, Counsell S (2005) Investigating web size metrics for early Web cost estimation. J Syst Softw 77(2):157–172
2. Jensen FV (1996) An introduction to Bayesian networks. UCL Press, London

Effort Prediction for Dynamic Web Applications Developed Using a Content Management System

9

Introduction

This chapter revisits the expert-based knowledge engineering of Bayesian networks process that was detailed in Chap. 6 (Fig. 9.1), describing the tasks carried out for each of the three main steps that form part of that process. Before starting the elicitation of the Dynamic Web applications effort estimation BN model, the domain experts participating were presented with an overview of Bayesian network models, and examples of "what-if" scenarios using a made-up BN. This, we believe, facilitated the entire process as the use of an example, and the brief explanation of each of the steps in the expert-based knowledge engineering of Bayesian networks process, provided a concrete understanding of what to expect. We also made it clear that the knowledge engineer was a facilitator of the process, and that the company's commitment was paramount for the success of the process.

The entire process took 54 person-hours to be completed, corresponding to nine 3-h slots.

The domain experts (DEs) who took part in this case study were project managers of a well-established Web company in Auckland (New Zealand). The company had ~20 employees, and branches overseas. The project managers had each worked in Web development for more than 10 years. In addition, this company developed a wide range of Web applications, from static and multimedia like to very large e-commerce solutions. They also used a wide range of Web technologies, thus enabling the development of Web 2.0 applications. Previous to using the BN model created, the effort estimates provided to clients deviated from actual effort within the range of 20–60 %.

E. Mendes, *Practitioner's Knowledge Representation*, DOI 10.1007/978-3-642-54157-5_9,   141
© Springer-Verlag Berlin Heidelberg 2014

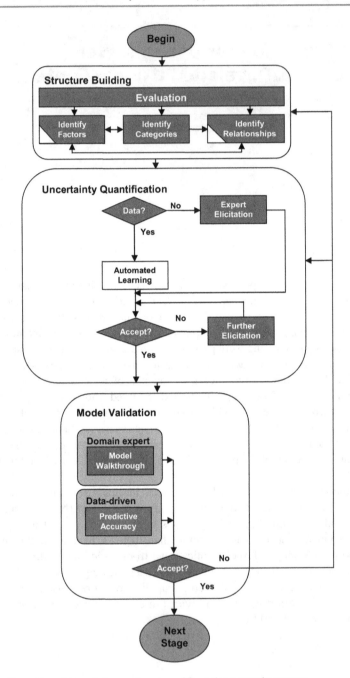

Fig. 9.1 Expert-based knowledge engineering of Bayesian networks process

Detailed Structure Building and Uncertainty Quantification

In order to identify the fundamental factors that the DEs took into account when preparing a project quote, we used the set of variables from the Tukutuku dataset [1] as a starting point (Table 9.1). We first sketched them out on a whiteboard, each one inside an oval shape, and then explained what each one meant within the context of the Tukutuku project. Our previous experience eliciting BNs in other domains (e.g., ecology) suggested that it was best to start with a few factors (even if they were not to be reused by the DE), rather than to use a "blank canvas" as a starting point.

Within the context of the Tukutuku project, a new high-effort feature/function requires at least 15 h to be developed by one experienced developer, and a high-effort adapted feature/function requires at least 4 h to be adapted by one experienced developer. These values are based on collected data.

Once the Tukutuku variables had been sketched out and explained, the next step was to remove all variables that were not relevant for the DEs, followed by adding to the whiteboard any additional variables (factors) suggested by them. We also documented descriptions for each of the factors suggested. Next, we identified the states that each factor would take. All states were discrete. Whenever a factor represented a measure of effort (e.g., total effort), we also documented the effort range corresponding to each state, to avoid any future ambiguity. For example, "low level 1" total effort corresponded to 4–10 person-hours, etc. Once all states were identified and documented, it was time to elicit the cause and effect relationships. As a starting point to this task we used a simple medical example from [2] (Fig. 9.2).

This example clearly introduces one of the most important points to consider when identifying cause and effect relationships—the timeline of events. If smoking is to be a cause of lung cancer, it is important that the cause precedes the effect. This may sound obvious with regard to the example used; however, it is our view that the use of this simple example significantly helped the DEs understand the notion of cause and effect, and how this related to software effort and risk-estimation and the BN being elicited.

Once the cause and effect relationships were identified, the DEs seemed happy with the BN's causal structure, and the work on eliciting the probabilities was initiated. All probabilities were created from scratch, and the probabilities elicitation took ~24 h. While entering the probabilities, the DEs decided to revisit the BN's causal structure after revisiting their effort estimation process; therefore a new iteration of the structure building and uncertainty quantification took place. The final BN causal structure is shown in Fig. 9.3. Note that Fig. 9.3 is not a BN based directly on Table 9.1. Here we present the BN using belief bars rather than labelled factors, so readers can see the probabilities that were elicited. Note that this BN corresponds to the current model being used by the Web company (also validated, to be detailed next). It contains 37 factors and 43 relationships identified by the 2 DEs as fundamental for Web effort estimation.

Each of the factors used in the Dynamic Web applications effort estimation BN model is presented in Table 9.2. Whenever a description is missing, this is because the factor's name is already self-explanatory.

Table 9.1 Tukutuku variables

	Variable name	Description
Project data	TypeProj	Type of project (new or enhancement)
	nLang	Number of different development languages used
	DocProc	If project followed defined and documented process
	ProImpr	If project team involved in a process improvement programme
	Metrics	If project team part of a software metrics programme
	DevTeam	Size of a project's development team
	TeamExp	Average team experience with the development language(s) employed
Web application	TotWP	Total number of Web pages (new and reused)
	NewWP	Total number of new Web pages
	TotImg	Total number of images (new and reused)
	NewImg	Total number of new images created
	Num_Fots	Number of features reused without any adaptation
	HFotsA	Number of reused high-effort features/functions adapted
	Hnew	Number of new high-effort features/functions
	TotHigh	Total number of high-effort features/functions
	Num_FotsA	Number of reused low-effort features adapted
	New	Number of new low-effort features/functions
	TotNHigh	Total number of low-effort features/functions

Fig. 9.2 A small BN illustrating cause and effect relationships

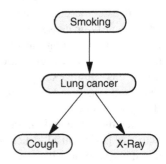

Detailed Model Validation

Both model walk-through and predictive accuracy were used to validate the Web effort BN model, where the former was the first type of validation to be employed. The DEs used four different scenarios to check whether the factor tot_effort would provide the highest probability to the effort state that corresponded to the DEs' own suggestions. All scenarios were run successfully; however, it was also necessary to use data from past projects, for which total effort was known, in order to check the model's calibration. A validation set containing data on 11 projects was used. The DEs selected a range of projects presenting different sizes and levels of complexity,

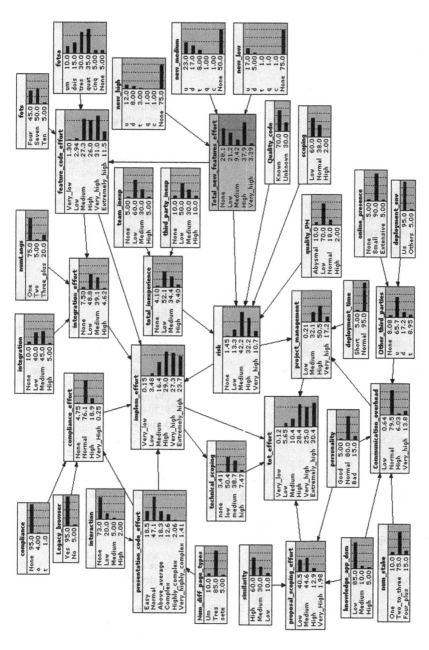

Fig. 9.3 Expert-based Web effort estimation Bayesian network model

Table 9.2 Description of factors identified by the domain experts

Factor	Categories	Description, observation
Number of languages used in the content	–1, 2, 3+	Real languages (e.g., English, Chinese), not programming languages
Client's personality	Good, normal, bad	Strong ideas, control freak, unfocused Description for each category: good (have done part of the planning even before the project starts, willing to listen to alternatives), normal (average, no real effect, not proactive but also doesn't hinder), bad (totally unfocused, inconsistent)
Client's knowledge of the application domain	Low, medium, high	Clear idea of what they want to achieve, what the application needs to do
Number of stakeholders involved	Single, low, high	Number of stakeholders involved—number of companies/people (client-side) involved in the process: single (1 person), low (2–3), high (4+)
Similarity to previous projects	High, medium, low	Similarity of domain/functionality/design
Quality of existing code being adapted/integrated	Known, unknown	Applies to both code developed in house, and to third party code
Number of features[a] off the shelf being used	(4–6), (7–9), (10+)	Here each feature requires a very low/low effort to be adapted (30 min up to 6, then 1 h total up to 9; 10+ would take 1½ h, average of 5 min per feature)
Number of features off the shelf being used that require adaptation	None, 1, 2, 3, 4, 5	Here each feature requires medium effort (~2 h) to be adapted.
Number of high-effort new features that need developing	None, 1, 2, 3, 4, 5	Here each feature requires ~15+ (more like 20 h) to be developed
Number of medium effort new features that need developing	None, 1, 2, 3, 4, 5	Here each feature requires ~10–15 (more like 12 h) to be developed
Number of low effort new features that need developing	None, 1, 2, 3, 4, 5	Here each feature requires ~5 h to be developed
Deployment environment	Us, others	If this company is hosting the Web application, or if a third party will get involved
Existing online presence	None, small, extensive	Existing domain names, email addresses, websites that the client already has

(continued)

Table 9.2 (continued)

Factor	Categories	Description, observation
Number of different page types	(1–2) (3–6) (7+)	Different page layouts
Amount of interaction in the application	None, low, medium, high	E.g., interaction in the application (to give immediate feedback on forms, how to present Google maps, etc.)
Level of integration	None, low, medium, high	Relates to the level of coupling (interaction) between features), and how much this will impact testing
Deployment time	Short, normal	If the client wants the site deployed quickly, generally results in more post-deployment work to optimise it
Quality of project management	Abysmal, low, normal, high	Degree of involvement of the project manager, and also their skills (inexperience)
Team inexperience	None, low, medium, high	Team's average experience with designing websites, experience with the languages used, experience with using the Web (browsing and awareness of what's possible)
Number of third parties involved	None, 1, 2, 3+	E.g., subcontractors, printing, SMS gateways, hosting providers, domain registration, payment providers
Third party inexperience	None, low, medium, high	E.g., subcontractors (including own designers), SMS gateways, hosting providers, domain registration, payment providers
Total inexperience	None, low, medium, high	Optimisation factor
Proposal scoping effort	Low (0+ to 1 person-hour), medium (1+ to 2 person-hours), high (2+ to 5 person-hours), very high (5+ person-hours)	Identify what the site is going to do—(technical requirements, marketing requirements (what the site owner will get out of it), user requirements (what the visitors will get out of it); should be a scope that complies with the available budget. No tangible specification document is generated; the scope is worked out inside the two project managers' minds
Technical scoping effort	None, low (2–5 % of the implementation effort), medium (5 % + to 7 % of the implementation effort), high (7 % + to 10 % of the implementation effort)	Identify how the site is going to do what it has to do (technical requirements); should be a scope that complies with the available budget. A tangible specification should be generated

(continued)

Table 9.2 (continued)

Factor	Categories	Description, observation
Level of technical scoping	Low, normal, high	Level of project planning, technical requirements
Legacy browser support	Yes, no	Ensure back compatibility with IE6 etc. If Yes, then it's 50 % of presentation code effort
Presentation code effort	Easy (4–6 person-hours), normal (6+ to 10 person-hours), above average (10+ to 15 person-hours), complex (15+ to 20 person-hours), highly complex (20+ to 30 person-hours), very highly complex (30+ person-hours)	Amount of effort to write html, javascript and css
Compliance effort	None, normal (0+ to 7.5 person-hours), high (7.5+ to 20 person-hours), very high (20+ person-hours)	
Compliance	None, o (1–50 % of presentation code effort), t (2–75 % of presentation code effort)	Government websites have to comply with standard accessibility guidelines/etc., accessibility
Risk Factor	None, low, medium, high, very high	Risk of increasing effort compared to the ideal effort
Total effort	Very low (4+ to 10 person-hours), low (10+ to 25 person-hours), medium (25+ to 40 person-hours), high (40+ to 80 person-hours), very high (80+ to 150 person-hours), extremely high (150+ person-hours)	
Implementation effort	Very low (4+ to 7 person-hours), low (7+ to 15 person-hours), medium (15+ to 30 person-hours), high (30+ to 60 person-hours), very high (60+ to 120 person-hours), extremely high (120+ person-hours)	
Project management factor	Low (10 % to 15 % of implementation effort), medium (15 + % to 20 % of implementation effort), high (20 + % to 25 % of implementation effort), very high (25 + % of implementation effort)	Also includes the planning of the application, and any training that needs to be done so staff can get up to speed
Integration effort	None, low (0+ to 2 person-hours), medium (2+ to 8 person-hours), high (8+ person-hours)	

(continued)

Table 9.2 (continued)

Factor	Categories	Description, observation
Feature code effort	Very low (0+ to 1 person-hour), low (1+ to 4 person-hours), medium (4+ to 12 person-hours), high (12+ to 30 person-hours), very high (30+ to 80 person-hours), extremely high (80+ person-hours)	
Total new features effort	None, low (5+ to 12 person-hours), medium (12+ to 25 person-hours), high (25+ to 80 person-hours), very high (80+ person-hours)	Optimisation factor
Communication overhead	Low overhead, normal, high overhead, very high overhead	Optimisation factor (not quantified as #person-hours)

[a]Note: Features apply to features developed in-house and also by third-parties. Features within this context mean functionality (here they include the testing to work on each feature, but not the integration testing looking at the coupling between features)

where all 11 projects were representative of the types of projects developed by the Web company: 5 were small projects; 2 were medium, 2 large, and 1 very large.

For each project, evidence was entered in the BN model, and the effort range corresponding to the highest probability provided for "tot_effort" was compared to that project's actual effort (see an example in Fig. 9.4). The company had also defined the range of effort values associated with each of the categories used to measure "tot_effort". In the case of the company described herein, medium effort corresponds to 25–40 person-hours. Whenever actual effort did not fall within the effort range associated with the category with the highest probability, there was a mismatch; this meant that some probabilities needed to be adjusted.

Whenever probabilities were adjusted, we re-entered the evidence for each of the projects in the validation set that had already been used in the validation step to ensure that the calibration already carried out had not been affected. This was done to ensure that each calibration would always be an improvement upon the previous one. Within the scope of the model presented herein, of the 11 projects used for validation, only 1 required the model to be recalibrated. This means that for all the ten projects remaining, the BN model presented the highest probability to the effort range that contained the actual effort for the project being used for validation. Once all 11 projects were used to validate the model the DEs assumed that the validation step was complete.

The BN model was completed in September 2009, and has been successfully used to estimate effort for new projects developed by the company. In addition, the two DEs changed their approach to estimating effort as follows: prior to using the BN model, these DEs had to elicit requirements using very short meetings with clients, given that these clients assumed that short meetings were enough in order to understand what the applications needed to provide once delivered. The DEs were also not fully aware of the factors that they subjectively took into account when preparing an effort estimate; therefore many times they ended up providing unrealistic estimates to clients.

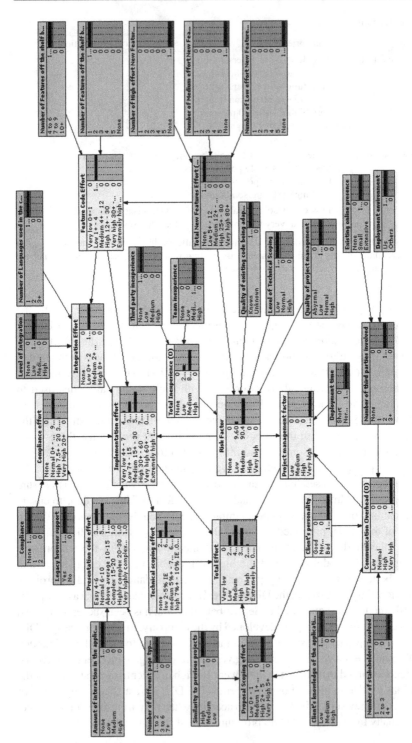

Fig. 9.4 Example of evidence being entered in the Web effort BN model

Once the BN model was validated, the DEs started to use the model not only to obtain better estimates than the ones previously prepared by subjective means, but also as means to guide their requirements elicitation meetings with prospective clients. They targeted their questions at obtaining evidence to be entered in the model as the requirements meetings took place; by doing so, they basically had effort estimates that were practically ready to use for costing the projects, even when meetings with clients had short durations. Such changes in approach provided extremely beneficial to the company, given that all estimates provided using the model turned out to be more accurate on average than the ones previously obtained by subjective means.

Clients were not presented the model due to its complexity; however, by entering evidence while a requirements elicitation meeting took place, the DEs were able to optimize their elicitation process by being focused and factor-driven.

We believe that the successful development of this Web effort BN model was greatly influenced by the commitment of the company, and also by the DEs' experience estimating effort.

Conclusions

This chapter has presented a case study where a Bayesian model for Web effort estimation was built using solely knowledge of two DEs from a well-established Web company in Auckland, New Zealand. This model was developed using an adaptation of the knowledge engineering for Bayesian networks process. Its causal structure went through three versions, because as the work progressed the experts' views on which factors were fundamental when they estimated effort also matured. Each session with the DEs lasted for no longer than 3 h. The final BN model was calibrated using data on the 11 past projects. These projects represented typical projects developed by the company, and were believed by the experts to provide enough data for model calibration.

Since the model's adoption, it has been successfully used to provide effort quotes for the new Web projects managed by the company.

The entire process used to build and validate the BN model took 54 person-hours, where the largest amount of time was spent eliciting the probabilities. This is an issue to those building BN models from domain expertise only, and is currently the focus of our future work.

The elicitation process enables experts to think deeply about their effort estimation process and the factors taken into account during that process, which in itself is advantageous to a company. This has been pointed out to us not only by the domain experts whose model is presented herein, but also by other companies with which we worked on model elicitations.

References

1. Mendes E, Mosley N, Counsell S (2005) Investigating Web size metrics for early Web cost estimation. J Syst Softw 77(2):157–172
2. Jensen FV (1996) An introduction to Bayesian networks. UCL Press, London

Effort Prediction to Manage Outsourcing Projects for the Development of Web Hypermedia and Web Software Applications

10

Introduction

This chapter revisits the expert-based knowledge engineering of Bayesian networks (EKEBN) process that was detailed in Chap. 6 (Fig. 10.1), describing the tasks carried out for each of the three main steps that form part of that process. Before starting the elicitation of the Web hypermedia and software effort Bayesian network model, the domain expert participating was presented with an overview of Bayesian network models, and examples of "what-if" scenarios using a made-up Bayesian network. This, we believe, facilitated the entire process as the use of an example, and the brief explanation of each of the steps in the EKEBN process, provided a concrete understanding of what to expect. We also made it clear that the knowledge engineer was a facilitator of the process, and that the Web company's commitment was paramount for the success of the process.

The entire process took 18 h to be completed, corresponding to 36 person-hours in six 3-h slots.

The domain expert (DE) who took part in this case study is the project manager (and owner) of a well-established Web company in Auckland (New Zealand). The company has one project manager, two developers employed by the company, and several subcontractors. The project manager has worked in Web development for more than 10 years (back in 2008), and his company develops a wide range of Web applications, from static and multimedia-like to very large e-commerce solutions. They also use a wide range of Web technologies, thus enabling the development of Web 2.0 applications. Previous to using the BN model created, the effort estimates provided to clients would deviate from actual effort within the range of 10–40 %.

E. Mendes, *Practitioner's Knowledge Representation*, DOI 10.1007/978-3-642-54157-5_10,   153
© Springer-Verlag Berlin Heidelberg 2014

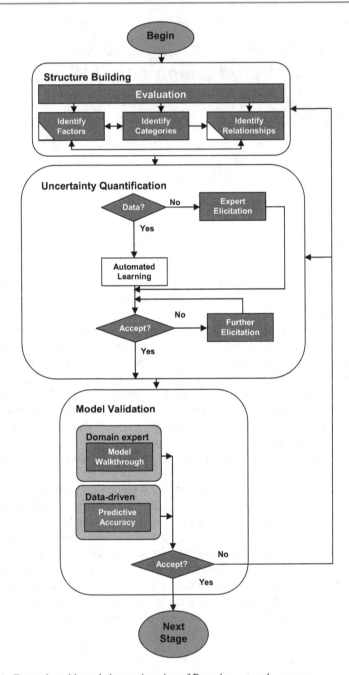

Fig. 10.1 Expert-based knowledge engineering of Bayesian networks process

Table 10.1 Tukutuku variables

	Variable name	Description
Project data	*TypeProj*	Type of project (new or enhancement)
	nLang	Number of different development languages used
	DocProc	If project followed defined and documented process
	ProImpr	If project team involved in a process improvement programme
	Metrics	If project team part of a software metrics programme
	DevTeam	Size of a project's development team
	TeamExp	Average team experience with the development language(s) employed
Web application	*TotWP*	Total number of Web pages (new and reused)
	NewWP	Total number of new Web pages
	TotImg	Total number of images (new and reused)
	NewImg	Total number of new images created
	Num_Fots	Number of features reused without any adaptation
	HFotsA	Number of reused high-effort features/functions adapted
	Hnew	Number of new high-effort features/functions
	TotHigh	Total number of high-effort features/functions
	Num_FotsA	Number of reused low-effort features adapted
	New	Number of new low-effort features/functions
	TotNHigh	Total number of low-effort features/functions

Detailed Structure Building and Uncertainty Quantification

In order to identify the fundamental factors that the DEs took into account when preparing a project quote, we used the set of variables from the Tukutuku dataset [1] as a starting point (Table 10.1). We first sketched them out on a whiteboard, each one inside an oval shape, and then explained what each one meant within the context of the Tukutuku project. Our previous experience eliciting BNs in other domains (e.g., ecology) suggested that it was best to start with a few factors (even if they were not to be reused by the DE), rather than to use a "blank canvas" as a starting point.

Within the context of the Tukutuku project, a new high-effort feature/function requires at least 15 h to be developed by one experienced developer, and a high-effort adapted feature/function requires at least 4 h to be adapted by one experienced developer. These values are based on collected data.

Once the Tukutuku variables had been sketched out and explained, the next step was to remove all variables that were not relevant for the DE, followed by adding to the whiteboard any additional variables (factors) suggested by them. We also documented descriptions for each of the factors suggested. Next, we identified the states that each factor would take. All states were discrete. Whenever a factor represented a measure of effort (e.g., total effort), we also documented the effort range corresponding to each state, to avoid any future ambiguity. For example,

Fig. 10.2 An example of cause and effect relationships

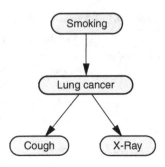

"very low" total effort corresponded to 0+ to 8 person-hours, etc. Once all states were identified and documented, it was time to elicit the cause and effect relationships. As a starting point to this task we used a simple medical example from [2] (Fig. 10.2).

This example clearly introduces one of the most important points to consider when identifying cause and effect relationships—the timeline of events. If smoking is to be a cause of lung cancer, it is important that the cause precedes the effect. This may sound obvious with regard to the example used; however, it is our view that the use of this simple example significantly helped the DEs understand the notion of cause and effect, and how this related to software effort and risk-estimation and the BN being elicited.

Once the cause and effect relationships were identified the Web effort estimation causal structure was as follows (Fig. 10.3). Note that Fig. 10.3 is not a BN based directly on Table 10.1.

The two factors "total effort" and "overall complexity" were each reached by a large number of relationships; therefore this structure needed to be simplified in order to reduce the number of probabilities to be elicited. New factors were suggested by the knowledge engineer (names ending in "_N"; see Fig. 10.4), and validated by the DE. Note that the extension _N, which stands for neutral, was chosen by the DE. The DE also made a few more changes to some of the relationships, leading to the BN causal structure presented in Fig. 10.4.

At this point the DE seemed happy with the BN's causal structure, and the work on eliciting the probabilities was initiated. All probabilities were created from scratch, a very time-consuming task (~8 to 10 h).

While entering the probabilities, the DE decided to revisit the BN's causal structure after revisiting the effort estimation process; therefore a new iteration of the structure building step took place. The final BN causal structure is shown in Fig. 10.5. Here we present the BN using belief bars rather than labelled factors, so readers can see the probabilities that were elicited. Note that this BN corresponds to the current model used by the Web company. This model contains 15 factors and 14 relationships identified by the DE as fundamental for Web effort estimation.

The description of each of the factors used in the Web hypermedia and software effort Bayesian network model is given in Table 10.2.

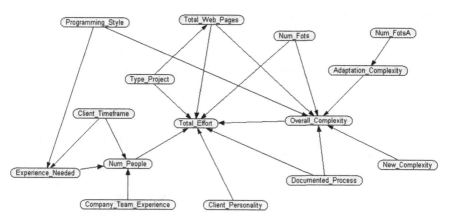

Fig. 10.3 First version of the model's causal structure

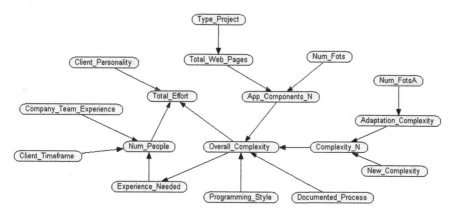

Fig. 10.4 An updated version of the model's causal structure

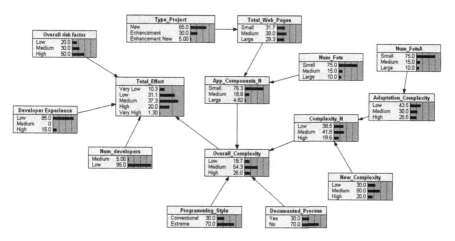

Fig. 10.5 Final version of the model's causal structure

Table 10.2 Description of factors elicited from the domain expert

Factor	Categories	Description, observation
Total_Web_Pages	Small, medium, large	Amount of Web pages
Overall risk factor	Low, medium, high	Is it a good customer? Easy going, average, difficult
Type_Project	New, enhancement, enhancement new	New, enhancement (developed by the company), enhancement and new (enhancement to a project that was developed by a third party)
Num_developers	Low, medium	Low (1 person), medium (2+)
Developer experience	Low, medium, high	Based on the number years if solo, or as a combination of skills if there is more than one developer
Programming_Style	Conventional, extreme	How well the client knows the requirements; conventional (means that the client knows the requirements well), extreme (the client doesn't know enough and keeps changing requirements)
Num_Fots	Small (0–5), medium (6–15), large (16+)	Total number of new features/functions being developed from scratch for the application
Num_FotsA	Small (0–5), medium (6–15), large (16+)	Total number of features/functions that are being reused with some level of adaptation
Adaptation_Complexity	Low, medium, high	Complexity associated with adapting a feature
New_complexity	Low, medium, high	Complexity associated with all the new features/functions to be developed from scratch
Documented_Process	Yes, no	Yes, no (they document most of the time, for large projects, at the start of the project so they have a plan, but also at the end of the project so it becomes easier to adapt features/functions on past projects)
Total_Effort	Very low (0+ to 8 person-hours), low (8+ to 25 person-hours), medium (25+ to 50 person-hours), high (50+ to 100 person-hours), very high (100+ person-hours)	
Overall_Complexity	Low, medium, high (effort)	
Complexity_N	Low, medium, high (effort)	
App_Components_N	Small, medium, large	Combined number of components

Detailed Model Validation

Both model walk-through and predictive accuracy were used to validate the Web effort estimation BN model, where the former was the first type of validation to be employed. The DE used four different scenarios to check whether the factor Total_Effort would provide the highest probability to the effort state that corresponded to the DE's own suggestion.

All scenarios were run successfully; however, it was also necessary to use data from past projects, for which total effort was known, in order to check the model's calibration.

A validation set containing data on eight projects was used. The DE selected a range of projects presenting different sizes and levels of complexity, where all eight projects were representative of the types of projects developed by the Web company: four were small projects; three were medium and one was large.

For each project, evidence was entered in the BN model (an example is given in Fig. 10.6b, where evidence is characterised by dark grey factors with probabilities equal to 100 % (1...)), and the effort range corresponding to the highest probability provided for "total effort" was compared to that project's actual effort. For example, in Fig. 10.6b, this would correspond to "total effort" = medium. The company had also defined the range of effort values associated with each of the categories used to measure "total estimated effort". In the case of the company described herein, medium effort corresponded to 25–50 person-hours. Whenever actual effort did not fall within the effort range associated with the category with the highest probability, there was a mismatch; this meant that some probabilities needed to be adjusted. However, within the context of this work, hardly any calibration was needed.

Whenever probabilities were adjusted, we re-entered the evidence for each of the projects in the validation set that had already been used in the validation step to ensure that the calibration already carried out had not been affected. This was done to ensure that each calibration would always be an improvement upon the previous one. Once all eight projects were used to calibrate the model, the domain expert assumed that the validation step was complete.

Figure 10.6 shows two scenarios of use for the Web effort estimation BN. The first (Fig. 10.6a) shows the likely probabilities for all factors in the BN given an expected Total_effort = very high (grey factor in Fig. 10.6a); conversely, the second scenario shows the likely probabilities for Total_effort when evidence is entered along the BN (grey factors).

This BN model has been in production since 2008 and has been successfully used to estimate effort for numerous projects. The domain expert uses solely the model to obtain effort estimates, rather than to combine their tacit knowledge of previous projects with the model's proposed effort estimate.

We believe that the successful development of this Web effort estimation Bayesian network model was greatly influenced by the commitment of the company, and also by the DE's exceptional experience estimating effort.

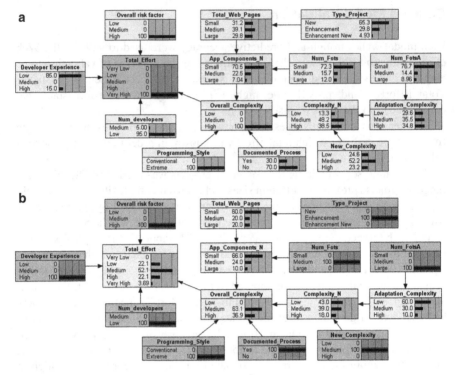

Fig. 10.6 Diagnostic and predictive scenarios using the Web effort BN model

---

Conclusions

This chapter has presented a case study where a Bayesian model for Web effort estimation was built using solely knowledge of one domain expert from a well-established Web company in Auckland, New Zealand. This model was developed using the expert-based knowledge engineering for Bayesian networks process (Fig. 10.1).

Each session with the DE lasted for no longer than 3 h. The final Bayesian network model was calibrated using data on eight past projects. These projects represented typical projects developed by the company, and believed by the experts to provide enough data for model calibration.

Since the model's adoption, it has been successfully used to provide effort quotes for the new projects managed by the company. The entire process used to build and validate the Bayesian network model took 36 person-hours.

The elicitation process enables experts to think deeply about their effort estimation process and the factors taken into account during that process, which in itself is advantageous to a company. This has been pointed out to us not only by the domain expert whose model is presented herein, but also by other companies with which we worked on model elicitations.

References

1. Mendes E, Mosley N, Counsell S (2005) Investigating web size metrics for early web cost estimation. J Syst Softw 77(2):157–172
2. Jensen FV (1996) An introduction to Bayesian networks. UCL Press, London

References

Effort Prediction for Game Applications Delivered on the Web

11

Introduction

This chapter revisits the expert-based knowledge engineering of Bayesian networks (EKEBN) process that was detailed in Chap. 6 (Fig. 11.1), describing the tasks carried out for each of the three main steps that form part of that process. Before starting the elicitation of the Web-based Games effort Bayesian network model, the domain expert (DE) participating was presented with an overview of Bayesian network models, and examples of "what-if" scenarios using a made-up Bayesian network. This, we believe, facilitated the entire process as the use of an example, and the brief explanation of each of the steps in the EKEBN process, provided a concrete understanding of what to expect. We also made it clear that the knowledge engineer was a facilitator of the process, and that the Web company's commitment was paramount for the success of the process.

The entire process took 126 person-hours to be completed, corresponding to twenty-one 3-h slots.

The domain expert who took part in this case study is the project manager (and owner) of a well-established Web company in Auckland (New Zealand). The company had five employees, and also outsourced work. The project manager had worked in Web development for more than 10 years. In addition, this company developed a wide range of Web software applications, from heavy multimedia-like to very large e-commerce solutions. They also used a wide range of Web technologies, thus enabling the development of Web 2.0 applications. Previous to using the BN model created, the effort estimates provided to clients deviated from actual effort within the range of 30–60 %.

E. Mendes, *Practitioner's Knowledge Representation*, DOI 10.1007/978-3-642-54157-5_11,   163
© Springer-Verlag Berlin Heidelberg 2014

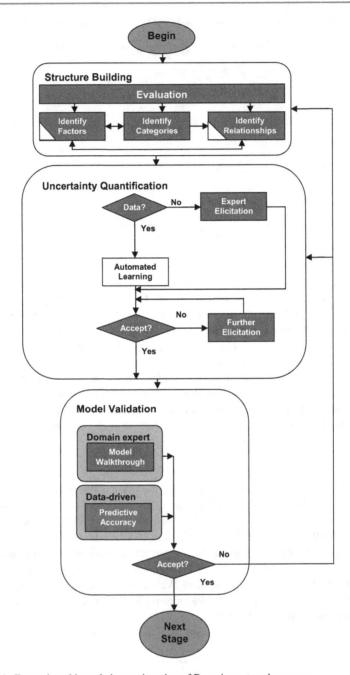

Fig. 11.1 Expert-based knowledge engineering of Bayesian networks process

Table 11.1 Tukutuku variables

	Variable name	Description
Project data	*TypeProj*	Type of project (new or enhancement)
	nLang	Number of different development languages used
	DocProc	If project followed defined and documented process
	ProImpr	If project team involved in a process improvement programme
	Metrics	If project team part of a software metrics programme
	DevTeam	Size of a project's development team
	TeamExp	Average team experience with the development language(s) employed
Web application	*TotWP*	Total number of Web pages (new and reused)
	NewWP	Total number of new Web pages
	TotImg	Total number of images (new and reused)
	NewImg	Total number of new images created
	Num_Fots	Number of features reused without any adaptation
	HFotsA	Number of reused high-effort features/functions adapted
	Hnew	Number of new high-effort features/functions
	TotHigh	Total number of high-effort features/functions
	Num_FotsA	Number of reused low-effort features adapted
	New	Number of new low-effort features/functions
	TotNHigh	Total number of low-effort features/functions

Detailed Structure Building and Uncertainty Quantification

In order to identify the fundamental factors that the DEs took into account when preparing a project quote, we used the set of variables from the Tukutuku dataset [1] as a starting point (Table 11.1). We first sketched them out on a whiteboard, each one inside an oval shape, and then explained what each one meant within the context of the Tukutuku project. Our previous experience eliciting BNs in other domains (e.g. ecology) suggested that it was best to start with a few factors (even if they were not to be reused by the DE), rather than to use a "blank canvas" as a starting point.

Once the Tukutuku variables had been sketched out and explained, the next step was to remove all variables that were not relevant for the DE, followed by adding to the whiteboard any additional variables (factors) suggested by the DE. We also documented descriptions for each of the factors suggested. Next, we identified the states that each factor would take. All states were discrete. Whenever a factor represented a measure of effort (e.g. total effort), we also documented the effort range corresponding to each state, to avoid any future ambiguity. For example, "extremely low" total effort corresponded to 0+ to 8 person hours, etc.

Within the context of the Tukutuku project, a new high-effort feature/function requires at least 15 h to be developed by one experienced developer, and a high-

Fig. 11.2 An example of
cause and effect relationships

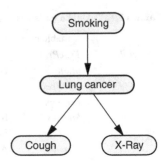

effort adapted feature/function requires at least 4 h to be adapted by one experienced developer. These values are based on collected data.

Once all states were identified and documented, it was time to elicit the cause and effect relationships. As a starting point to this task we used a simple medical example from [2] (Fig. 11.2).

This example clearly introduces one of the most important points to consider when identifying cause and effect relationships—the timeline of events. If smoking is to be a cause of lung cancer, it is important that the cause precedes the effect. This may sound obvious with regard to the example used; however, it is our view that the use of this simple example significantly helped the DE understand the notion of cause and effect, and how this related to Web effort estimation and the BN being elicited.

Once the cause and effect relationships were identified, the Web effort estimation causal structure was as follows (Fig. 11.3). Note that Fig. 11.3 is not a BN based directly on Table 10.1; however, it contains many of the factors that were part of the Tukutuku database.

The two factors "Total_Effort" and "Quality_Control" were each reached by a large number of relationships; therefore this structure needed to be simplified in order to reduce the number of probabilities to be elicited. New factors were suggested by the knowledge engineer (names ending in "_O", see Fig. 11.4), and validated by the domain expert. The DE also made a few more changes to some of the relationships, leading to the BN causal structure presented in Fig. 11.4.

At this point the DE seemed happy with the BN's causal structure and the work on eliciting the probabilities was initiated. While entering the probabilities, the DE decided to revisit the BNs causal structure; therefore a new iteration of the structure building step took place. The final BN causal structure is shown in Fig. 11.5. This same model is also shown in Fig. 11.6 using belief bars rather than labelled factors, so readers can see the probabilities that were elicited. Note that this model is the one used by the Web company. It contains 16 factors and 16 relationships identified by the domain expert as fundamental for Web effort estimation.

The description of each of the factors used in the Web-based Games effort Bayesian network model is given in Table 11.2.

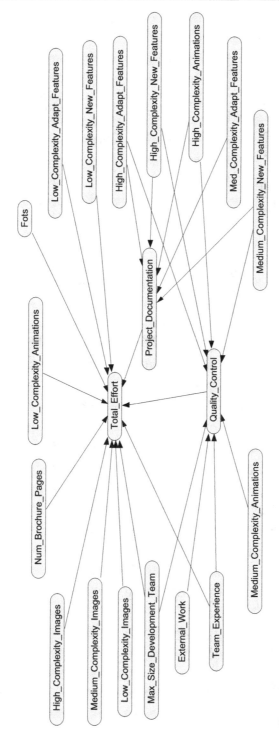

Fig. 11.3 First version of the model's causal structure

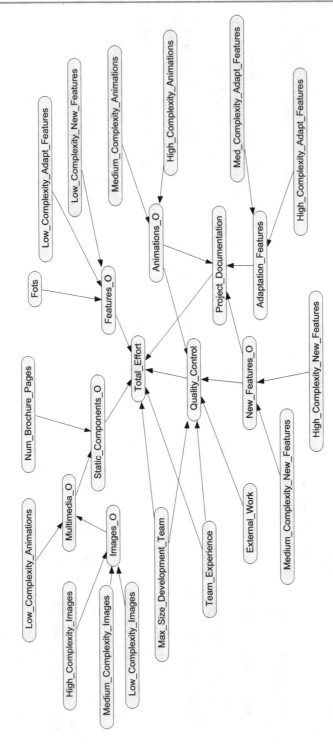

Fig. 11.4 An updated version of the model's causal structure

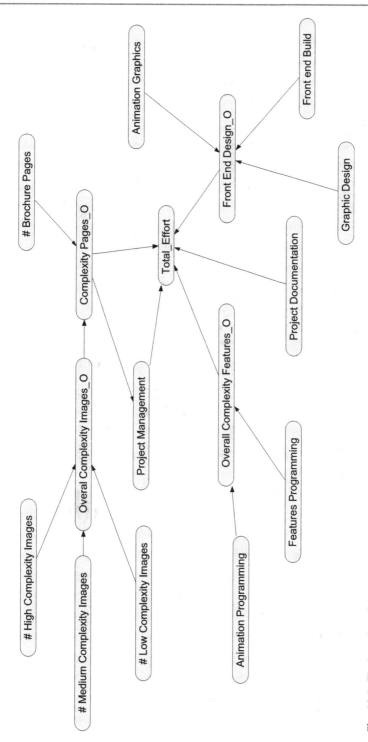

Fig. 11.5 Final version of the model's causal structure

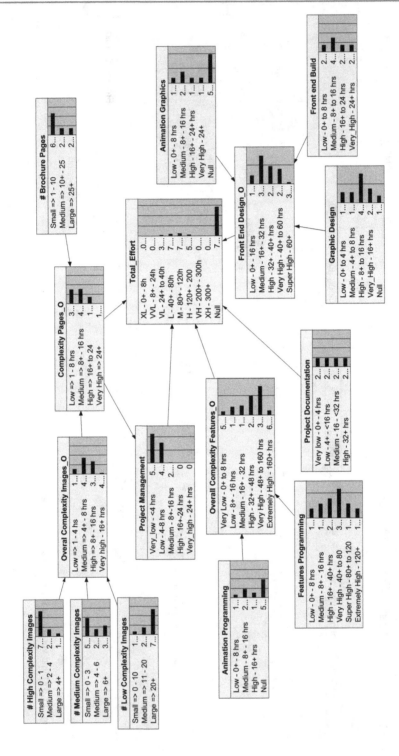

Fig. 11.6 Final Web effort estimation model

Table 11.2 Description of all the factors elicited from the domain expert

Factor	Categories	Description, observation
Graphic design	Low (0+ to 4), medium (4+ to 8), high (8+ to 16), very high (16+)	Effort to develop the interface, look and feel of the interface, mock-up of the interface
Front-end build	low—0+ to 8 person-hours; medium—8+ to 16 person-hours; high—16+ to 24 person-hours; very high—24+ person-hours	Effort to translate the graphic design into html
Front end Design_O	Low—0+ to 16 person-hours; medium—16+ to 32 person-hours; high—32+ to 40 person-hours; very high—40+ to 60 person-hours; super high—60+ person-hours	
Animation graphics	Low—0+ to 8 person-hours; medium—8+ to 16 person-hours; high—16+ to 24 person-hours; very high—24+ person-hours	Effort working on animations using an animation software
High-complexity images	Small—0–1 images; medium—2–4 images; large—4+ images	Herein an image takes between 30+ min and 2 h to be created
Medium-complexity images	Small—0–3 images; medium—4–6 images; large—6+ images	Herein an image takes between 10 min and 30 min to be created
Low-complexity images	Small—0–10 images; medium—11–20 images; large—20+ images	Herein an image takes between 0+ and 10 min to be created
Num brochure pages	Small—1–10 brochure pages; medium—11–25 brochure pages; large—25+ brochure pages	Counting the number of pages that are of type brochure. A brochure page takes around 15 min to be created
Features programming	Low—0+ to 8 person-hours; medium—8+ to 16 person-hours; high—16+ to 40 person-hours; very high—40+ to 80 person-hours; super high—80+ to 120 person-hours; extremely high—120+ person-hours	The amount of effort needed to implement (program) the features that the Web application will have
Overall Complexity Features_O	Very low—0+ to 8 person-hours; low—8+ to 16 person-hours; medium—16+ to 32 person-hours; high—32+ to 48 person-hours; very high—48+ to 160 person-hours; extremely high—160+ person-hours	The joint amount of effort needed to implement (program) the features and animations that the Web application will have
Project Documentation	Very low—0+ to 4 person-hours; low—4+ to 16 person-hours; medium—16+ to 32 person-hours; high—32+ person-hours	Effort that refers to the amount of documentation in terms of specification documents, user manuals, etc., that needs to be generated
Project management	Very low—0+ to 4 person-hours; low—4+ to 8 person-hours; medium—8+ to 16 person-hours; high—16+ to	Aspects to consider: quality control, client communication, follow up with clients, project deadlines

(continued)

Table 11.2 (continued)

Factor	Categories	Description, observation
	24 person-hours; very high—24+ person-hours	
Complexity pages_O	Low—1+ to 8 person-hours; medium— 8+ to 16 person-hours; high—16+ to 24 person-hours; very high—24+ person-hours	The joint amount of effort needed to construct the Web pages
Overall Complexity Images_O	Low—1+ to 4 person-hours; medium— 4+ to 8 person-hours; high—8+ to 16 person-hours; very high—16+ person-hours	The joint amount of effort needed to create the images to be used in the Web application
Total effort	Extremely low—0+ to 8 person-hours; very very low—8+ to 24 person-hours; very low—24+ to 40 person-hours; low—40+ to 80 person-hours; medium—80+ to 120 person-hours; high—120+ to 200 person-hours; very high—200+ to 300 person-hours; extremely high—300+ to person-hours	
Animation programming	Low—0+ to 8 person-hours; medium— 8+ to 16 person-hours; high—16+ person-hours	Effort working on animations by programming their movement

Detailed Model Validation

Both model walk-through and predictive accuracy were used to validate the Web effort estimation BN model, where the former was the first type of validation to be employed. The DE used six different scenarios to check whether the factor Total_Effort would provide the highest probability to the effort state that corresponded to the DE's own suggestion. All scenarios run successfully; however, it was also necessary to use data from past projects, for which total effort was known, in order to check the model's calibration.

A validation set containing data on 22 projects was used. The DE selected a range of projects presenting different sizes and levels of complexity, where all 22 projects were representative of the types of projects developed by the Web company.

For each project, evidence was entered in the BN model (an example is given in Fig. 11.7, where evidence is characterised by dark grey factors with probabilities equal to 100 % (1...)), and the effort range corresponding to the highest probability provided for "Total_Effort" was compared to that project's actual effort. For example, in Fig. 11.7, this would correspond to "Total_Effort" = VL (very large). The company had also defined the range of effort values associated with each of the

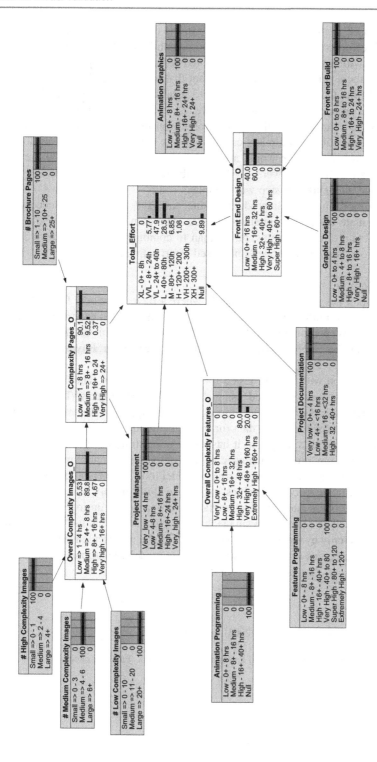

Fig. 11.7 Predictive scenario using the Web effort BN model

categories used to measure "Total_Effort". In the case of the company described herein, VL effort corresponded to 24+ to 40 person-hours. Whenever actual effort did not fall within the effort range associated with the category with the highest probability, there was a mismatch; this meant that some probabilities needed to be adjusted. Within the context of this work, 15 recalibrations were needed.

Whenever probabilities were adjusted, we re-entered the evidence for each of the projects in the validation set that had already been used in the validation step to ensure that the calibration already carried out was not affected. This was done to ensure that each calibration would always be an improvement upon the previous one. Once all 22 projects were used to calibrate the model the domain expert assumed that the validation step was complete.

This BN model has been in production since late 2009 and has been successfully used to estimate effort for numerous projects. The domain expert uses solely the model to obtain effort estimates, rather than to combine their tacit knowledge of previous projects with the model's proposed effort estimate.

Conclusions

This chapter has presented a case study where a Bayesian model for Web effort estimation was built using solely knowledge of one domain expert from a well-established Web company in Auckland, New Zealand. This model was developed using the Expert-based knowledge engineering for Bayesian networks process (Fig. 11.1).

Each session with the DE lasted for no longer than 3 h. The final Bayesian network model was calibrated using data on 22 past projects. These projects represented typical projects developed by the company, and were believed by the expert to provide enough data for model calibration.

Since the model's adoption, it has been successfully used to provide effort quotes for the new projects managed by the company. The entire process used to build and validate the Bayesian network model took 126 person hours.

The elicitation process enables experts to think deeply about their effort estimation process and the factors taken into account during that process, which in itself is already advantageous to a company. This has been pointed out to us not only by the domain expert whose model is presented herein, but also by other companies with which we worked on model elicitations.

References

1. Mendes E, Mosley N, Counsell S (2005) Investigating web size metrics for early web cost estimation. J Syst Softw 77(2):157–172
2. Jensen FV (1996) An introduction to Bayesian networks. UCL Press, London

Effort Prediction for Static and Dynamic Web Applications

12

Introduction

This chapter revisits the expert-based knowledge engineering of Bayesian networks (EKEBN) process that was detailed in Chap. 6 (Fig. 12.1), describing the tasks carried out for each of the three main steps that form part of that process. Before starting the elicitation of the Web static and dynamic effort Bayesian network model, the domain expert participating was presented with an overview of Bayesian network models and examples of "what-if" scenarios using a made-up Bayesian network. This, we believe, facilitated the entire process as the use of an example, and the brief explanation of each of the steps in the EKEBN process, provided a concrete understanding of what to expect. We also made it clear that the knowledge engineer was a facilitator of the process, and that the Web company's commitment was paramount for the success of the process.

The entire process took 120 person-hours to be completed, corresponding to twenty 3-h slots.

The domain expert (DE) who took part in this case study is the project manager of a well-established medium-size software company in Rio de Janeiro (Brazil). The company had ~30 employees working in Web projects. The project manager had worked in Web development for more than 10 years. In addition, this company developed a wide range of Web software applications using a content management system. Previous to using the BN model created, the effort estimates provided to clients would deviate from actual effort within the range of 40–50 %.

Detailed Structure Building and Uncertainty Quantification

In order to identify the fundamental factors that the DE took into account when preparing a project quote, we used the set of variables from the Tukutuku dataset [1] as a starting point (Table 12.1). We first sketched them out on a white-board, each one inside an oval shape, and then explained what each one meant within the

E. Mendes, *Practitioner's Knowledge Representation*, DOI 10.1007/978-3-642-54157-5_12, 175
© Springer-Verlag Berlin Heidelberg 2014

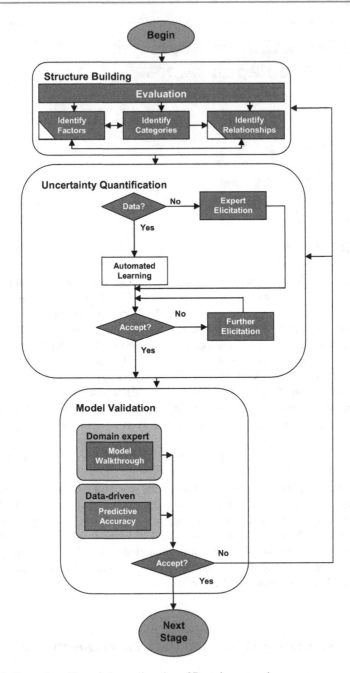

Fig. 12.1 Expert-based knowledge engineering of Bayesian networks process

Table 12.1 Tukutuku variables

	Variable name	Description
Project data	TypeProj	Type of project (new or enhancement)
	nLang	Number of different development languages used
	DocProc	If project followed defined and documented process
	ProImpr	If project team involved in a process improvement programme
	Metrics	If project team part of a software metrics programme
	DevTeam	Size of a project's development team
	TeamExp	Average team experience with the development language(s) employed
Web application	TotWP	Total number of Web pages (new and reused)
	NewWP	Total number of new Web pages
	TotImg	Total number of images (new and reused)
	NewImg	Total number of new images created
	Num_Fots	Number of features reused without any adaptation
	HFotsA	Number of reused high-effort features/functions adapted
	Hnew	Number of new high-effort features/functions
	TotHigh	Total number of high-effort features/functions
	Num_FotsA	Number of reused low-effort features adapted
	New	Number of new low-effort features/functions
	TotNHigh	Total number of low-effort features/functions

context of the Tukutuku project. Our previous experience eliciting BNs in other domains (e.g. ecology) suggested that it was best to start with a few factors (even if they were not to be reused by the DE), rather than to use a "blank canvas" as a starting point.

Once the Tukutuku variables had been sketched out and explained, the next step was to remove all variables that were not relevant for the DE, followed by adding to the whiteboard any additional variables (factors) suggested by them. We also documented descriptions for each of the factors suggested. Next, we identified the states that each factor would take. All states were discrete. Whenever a factor represented a measure of effort (e.g., total effort), we also documented the effort range corresponding to each state, to avoid any future ambiguity. For example, "maximum allocation/complex application" total effort corresponded to 801+ to 1,120 person-hours, etc.

Within the context of the Tukutuku project, a new high-effort feature/function requires at least 15 h to be developed by one experienced developer, and a high-effort adapted feature/function requires at least 4 h to be adapted by one experienced developer. These values are based on collected data.

Once all states were identified and documented, it was time to elicit the cause and effect relationships. As a starting point to this task we used a simple medical example from [2] (Fig. 12.2).

This example clearly introduces one of the most important points to consider when identifying cause and effect relationships—the timeline of events. If smoking

Fig. 12.2 An example of
cause and effect relationships

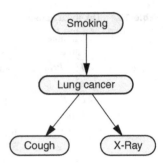

is to be a cause of lung cancer, it is important that the cause precedes the effect. This may sound obvious with regard to the example used; however, it is our view that the use of this simple example significantly helped the DE understand the notion of cause and effect, and how this related to Web effort estimation and the BN being elicited.

Once the cause and effect relationships were identified, the Web effort estimation causal structure was as follows (Fig. 12.3). Note that Fig. 12.3 is not a BN based directly on Table 12.1; however, it contains many of the factors that were part of the Tukutuku database. Note that the English translation of the factors' names is given in Fig. 12.4 and also in Table 12.2.

The final Bayesian network model contains 19 factors and 37 causal relationships identified by the domain expert as fundamental for Web effort estimation. In addition, it differs from all the previous models in that it did not include any made-up factors, i.e., factors that are not part of the original Bayesian network structure but that are included in order to reduce the amount of probabilities to elicit for a given factor.

This same model is also shown in Fig. 12.6 using belief bars rather than labelled factors, so readers can see the probabilities that were elicited.

The description of each of the factors used in the Web static and dynamic effort estimation BN model is given in Table 12.2.

Detailed Model Validation

Both model walk-through and predictive accuracy were used to validate the Web effort estimation BN model, where the former was the first type of validation to be employed. The DE used seven different scenarios to check whether the factor "total development effort" would provide the highest probability to the effort state that corresponded to the DE's own suggestion. All scenarios run successfully; however,

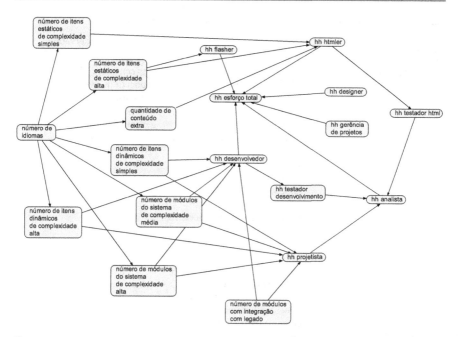

Fig. 12.3 Model's structure

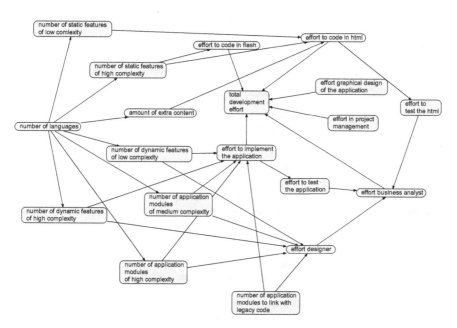

Fig. 12.4 Model's structure in English

Table 12.2 Description of all the factors elicited from the domain expert

Factor	Categories	Description, observation
Total development effort (hh esforço total)	Very small effort (alocacao muito pequena) (0+ to 32 person-hours) Small effort (alocacao pequena) (32+ to 60 person-hours) Effort small website (alocacao site pequeno) (61+ to 120 person-hours) Effort medium website (alocacao site medio)(120+ to 200 person-hours) Effort large website (alocacao site grande) (200+ to 320 person-hours) Maximum effort website (alocacao maxima site) (320+ to 640 person-hours) Effort simple/common system (alocacao sistema simples/comum) (640+ to 800 person-hours) Maximum effort complex system (alocacao maxima sistema complexo) (800+ to 1,120 person-hours)	This company differentiates between static Web applications, which they call websites, and Web applications with part of the content generated dynamically, which they call a system
Effort graphical design of the application (hh designer)	Help (apoio) (0+ to 5 person-hours) Banner (banner) (5+ to 10 person-hours) Hot site (hot site) (10+ to 20 person-hours) Website (site) (20+ to 50 person-hours) Complex website (site complicado) (50+ person-hours)	
Effort to code in flash (hh Flash)	Very small effort (alocacao muito pequena) (0 + to 4 person-hours) Small allocation (alocacao pequena) (4+ to 9 person-hours) Reasonable effort (alocacao razoavel) (9+ to 15 person-hours) Considerable effort (alocacao consideravel) (15+ to 25 person-hours) High effort (alocacao grande)	

(continued)

Table 12.2 (continued)

Factor	Categories	Description, observation
	(25+ to 40 person-hours) Very high effort (alocacao muito grande) (40+ to 70 person-hours) Abnormal (anormal) (70+ to 100 person-hours) Absurd (absurdo) (100+ person-hours)	
Effort to code in html (hh htmler)	Very small effort (alocacao muito pequena) (0+ to 4 person-hours) Small allocation (alocacao pequena) (4+ to 9 person-hours) Reasonable effort (alocacao razoavel) (9+ to 15 person-hours) Considerable effort (alocacao consideravel) (15+ to 25 person-hours) High effort (alocacao grande) (25+ to 40 person-hours) Very high effort (alocacao muito grande) (40+ to 70 person-hours) Abnormal (anormal) (70+ to 100 person-hours) Absurd (absurdo) (100+ person-hours)	Effort working on animations using an animation software
Effort business analyst (hh analista)	Zero (zero) (0 person-hours) Minimum effort (alocacao minima) (0+ to 4 person-hours) Effort small website (alocacao site pequeno) (4+ to 9 person-hours) Effort medium website (alocacao site medio) (9+ to 20 person-hours) Maximum effort website (alocacao maxima site) (20+ to 45 person-hours) Maximum effort simple system (alocacao maxima sistema simples) (45+ to 60 person-hours) Maximum effort complex system (alocacao maxima sistema complexo) (60+ to 75 person-hours)	This company differentiates between static Web applications, which they call websites, and Web applications with part of the content generated dynamically, which they call a system

(continued)

Table 12.2 (continued)

Factor	Categories	Description, observation
Effort to test the html (hh testador html)	Typical effort (alocacao típica) (0+ to 4 person-hours) High effort (alocacao grande) (4+ person-hours)	Here to test means to check the quality of the html that was written
Effort to test the application (hh testador desenvolvimento)	Zero (zero) (0 person-hours) Minimum effort (alocacao minima) (0+ to 4 person-hours) Effort small website (alocacao site pequeno) (4+ to 9 person-hours) Effort medium website (alocacao site medio) (9+ to 20 person-hours) Maximum effort website (alocacao maxima site) (20+ to 45 person-hours) Maximum effort simple system (alocacao maxima sistema simples) (45+ to 60 person-hours) Maximum effort complex system (alocacao maxima sistema complexo) (60+ to 75 person-hours)	Herein to test means to check whether the application has good quality (e.g., usability). This company differentiates between static Web applications, which they call websites, and Web applications with part of the content generated dynamically, which they call a system
Effort designer (hh projetista)	Very small effort (alocacao muito pequena) (0+ to 8 person-hours) Small effort (alocacao pequena) (8+ to 16 person-hours) Maximum effort website (alocacao maxima site) (16+ to 40 person-hours) Effort very small system (alocacao sistema muito pequeno) (40+ to 80 person-hours) Effort small system (alocacao sistema pequeno) (80+ to 120 person-hours) Effort usual system (alocacao sistema comum) (120+ to 240 person-hours) Effort complex system (alocacao sistema complexo) (240+ person-hours)	This company differentiates between static Web applications, which they call websites, and Web applications with part of the content generated dynamically, which they call a system
Number of application modules to link with legacy code (num modulos com integracao com legado)	Zero (zero) (0 person-hours) Very small (muito pequeno) (1 application module) Small (pequeno) (2 application	Modules here represent the code that embeds the business logic. It complements the code that is partially generated, and

(continued)

Table 12.2 (continued)

Factor	Categories	Description, observation
	modules) Medium (medio) (3 application modules) Large (grande) (4 application modules) Very large (muito grande) (5+ application modules)	also includes the code that manipulates the data stored on a database
Number of application modules of high complexity (num modulos do sistema de complexidade alta)	Zero (zero) (0 application modules) Very small (muito pequeno) (1–2 application modules) Small (pequeno) (3–4 application modules) Medium (medio) (5–6 application modules) Large (grande) (7–8 application modules) Very large (muito grande) (9–10 application modules)	Modules here represent the code that embeds the business logic. It complements the code that is partially generated, and also includes the code that manipulates the data stored on a database
Number of application modules of medium complexity (num modulos do sistema de complexidade media)	Zero (zero) (0 person-hours) Small (pequeno) (1–2 application modules) Unlikely (improvavel) (3–11 application modules) Expected (normal) (12+ application modules)	Modules here represent the code that embeds the business logic. It complements the code that is partially generated, and also includes the code that manipulates the data stored on a database
Number of dynamic features of low complexity (num itens dinamicos de complexidade simples)	Zero (zero) (0 dynamic features) Very small (muito pequeno) (1–2 dynamic features) Small (pequeno) (3–5 dynamic features) Medium (medio) (6–8 dynamic features) High (grande) (9–11 dynamic features) Very high (muito grande) (12+ dynamic features)	A dynamic feature represents any sort of animation that is provided in the application (e.g. flash animation)
Number of dynamic features of high complexity (num itens dinamicos de complexidade alta)	Zero (zero) (0 dynamic features) Very small (muito pequeno) (1–2 dynamic features) Small (pequeno) (3–5 dynamic features) Medium (medio) (6–8 dynamic features) High (grande) (9–11 dynamic features) Very high (muito grande) (12+ dynamic features)	

(continued)

Table 12.2 (continued)

Factor	Categories	Description, observation
Amount of extra content (quantidade de conteudo extra)	Zero (zero) (0 extra pages) Large (grande) (0+ to 4 extra pages) Very large (muito grande) (5–6 extra pages) Absurd (absurdo) (7+ extra pages)	Content herein represents pages of text
Number of static features of high complexity (num itens estaticos de complexidade alta)	Zero (zero) (0 static features) Very small (muito pequeno) (1–2 static features) Small (pequeno) (3–5 static features) Medium (medio) (6–8 static features) High (grande) (9–11 static features) Very high (muito grande) (12+ static features)	Features that do not present any animation (e.g., photo album, images)
Number of static features of low complexity (num itens estaticos de complexidade baixa)	Zero (zero) (0 static features) Very small (muito pequeno) (1–5 static features) Small (pequeno) (6–9 static features) Medium (medio) (10–25 static features) High (grande) (26–30 static features) Very high (muito grande) (31+ static features)	Features that do not present any animation (e.g., photo album, images)
Number of languages (num idiomas)	1 2 & 3 4	
Effort to implement the application (hh desenvolvedor)	Zero (zero) (0 person-hours) Very small effort (alocacao muito pequena) (0+ to 12 person-hours) Small effort (alocacao pequena) (12+ to 24 person-hours) Effort small website (alocacao site pequeno) (24+ to 60 person-hours) Effort medium website (alocacao site medio) (60+ to 130 person-hours) Maximum effort website (alocacao maxima site) (130+ to 300 person-hours) Maximum effort simple system (alocacao maxima sistema	This company differentiates between static Web applications, which they call websites, and Web applications with part of the content generated dynamically, which they call a system

(continued)

Table 12.2 (continued)

Factor	Categories	Description, observation
	simples) (300+ to 400 person-hours) Maximum allocation complex system (alocacao maxima sistema complexo) (400+ to 500 person-hours)	
Effort in project management (hh gerencia projetos)	10 (1 month) (1 mes) 20 (2 months) (2 meses) 30 (3 months) (3 meses) 40 (4 months) (4 meses) 50 (5 months) (5 meses) 60 (6 months) (6 meses) 80 (complex projects) (projetos complexos)	10 h for each month for which a project lasts

it was also necessary to use data from past projects, for which total effort was known, in order to check the model's calibration (Fig. 12.5).

A validation set containing data on nine projects was used. The DE selected a range of projects presenting different sizes and levels of complexity, where all nine projects were representative of the types of projects developed by the Web company. For each project, evidence was entered in the BN model (an example is given in Fig. 12.6, where evidence is characterised by dark grey factors with probabilities equal to 100 % (1...)), and the effort range corresponding to the highest probability provided for "total development effort" was compared to that project's actual effort. For example, in Fig. 12.6, this would correspond to "total development effort" = small (alocacao pequena). The company had also defined the range of effort values associated with each of the categories used to measure "total development effort". In the case of the company described herein, small effort corresponded to 33+ to 60 person-hours.

Whenever actual effort did not fall within the effort range associated with the category with the highest probability, there was a mismatch; this meant that some probabilities needed to be adjusted. Within the context of this work, five recalibrations were needed.

Whenever probabilities were adjusted, we re-entered the evidence for each of the projects in the validation set that had already been used in the validation step to ensure that the calibration already carried out was not affected. This was done to ensure that each calibration would always be an improvement upon the previous one. Once all nine projects were used to calibrate the model, the domain expert assumed that the validation step was complete.

This BN model has been in production since July 2011 and has been successfully used to estimate effort for numerous projects. The domain expert uses solely the model to obtain effort estimates, rather than to combine their tacit knowledge of previous projects with the model's proposed effort estimate.

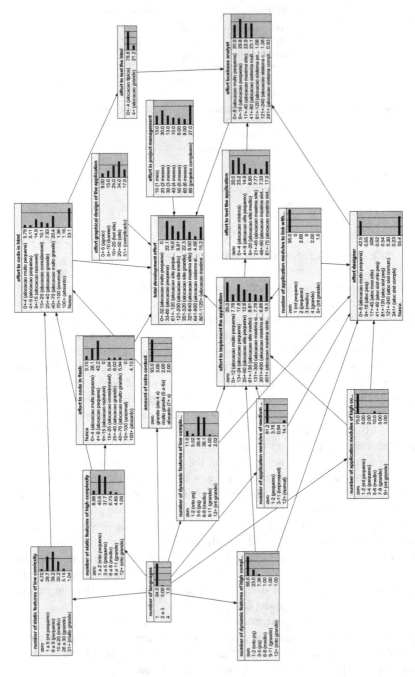

Fig. 12.5 Web effort estimation Bayesian network model (factors' names in English; categories in Portuguese)

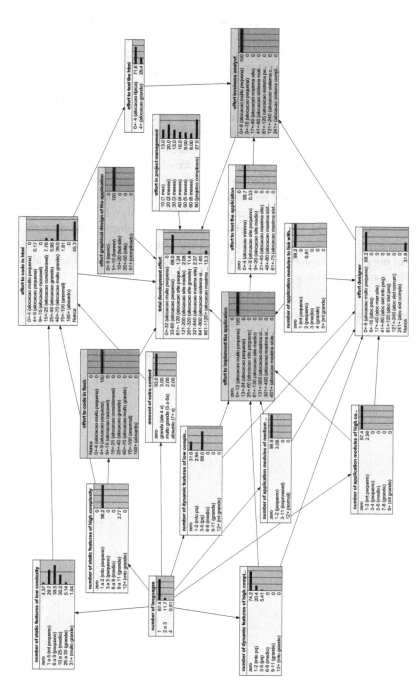

Fig. 12.6 Predictive scenario using the Web effort BN model (factors' names in English; categories in Portuguese)

Conclusions

This chapter has presented a case study where a Bayesian model for Web effort estimation was built using solely knowledge of one domain expert from a well-established Web company in Rio de Janeiro, Brazil. This model was developed using the expert-based knowledge engineering for Bayesian networks process (Fig. 12.1).

Each session with the DE lasted for no longer than 3 h. The final Bayesian network model was calibrated using data on nine past projects. These projects represented typical projects developed by the company, and believed by the expert to provide enough data for model calibration.

Since the model's adoption, it has been successfully used to provide effort quotes for the new projects managed by the company. The entire process used to build and validate the Bayesian network model took 120 person-hours.

The elicitation process enables experts to think deeply about their effort estimation process and the factors taken into account during that process, which in itself is already advantageous to a company. This has been pointed out to us not only by the domain expert whose model is presented herein, but also by other companies with which we worked on model elicitations.

References

1. Mendes E, Mosley N, Counsell S (2005) Investigating web size metrics for early web cost estimation. J Syst Softw 77(2):157–172
2. Jensen FV (1996) An introduction to Bayesian networks. UCL Press, London

Ways in Which to Use Bayesian Network Models Within a Company

13

Introduction

This book has detailed in Chaps. 7–12 how six different expert-based effort estimation BN models were built for several software companies in New Zealand and Brazil. Each of these companies employed their BN models in different ways, thus providing a wide range of scenarios of use that, in our view, can also be useful to other companies that wish to build and employ such models. This chapter therefore presents suggestions of how such estimation models can be employed, which are based on the scenarios implemented by the participating companies. Note that our suggestions are based on post-mortem meetings with the companies.

Using BNs as Part of a Wider Strategy for a Learning Organisation

Once a BN model has been built and validated, it is important that it does not remain within the boundaries of just a single development team and project manager (assuming the company has several development teams and project managers).

Our anecdotal evidence from post-mortem interviews with some of the companies with whom we collaborated building such models provided us with a range of concrete and industry-informed choices that will be detailed next. We believe that such suggestions can also be beneficial to other companies who are willing to use such models effectively as part of wider learning organisation strategies:

Process Improvement

A company can engage in improving their current effort estimation process using the steps detailed next:

E. Mendes, *Practitioner's Knowledge Representation*, DOI 10.1007/978-3-642-54157-5_13,   189
© Springer-Verlag Berlin Heidelberg 2014

- Presenting a seminar to all developers and project managers who participate in the development and management of the types of applications that were the focus of the expert-based model built. One of the goals of this seminar is to elaborate on the value that the entire company can gain from using such models. We suggest that the seminar focuses on presenting and detailing the model and various "what-if" scenarios based on their most recent projects for which effort was estimated using the model. This provides concrete examples of how the model is being used in the company. It is, in our view, also important to detail all the factors and categories that were defined by the domain experts who participated in building the model, such that all those attending the seminar become familiar with the terminology.
- Once the seminar takes place, the documentation relating to the model (description of factors and how to use the model) should become available for all the participants. The tool that is used to run the model and the model itself should also be made available to all development teams, so they can all run "what-if" scenarios using the model, as means for decision making relating to all effort estimates that need to be prepared.
- The nomenclature that was defined in the model should also become a common vocabulary for all teams, to be used whenever they need to discuss anything relating to effort estimates. This is quite important as it guarantees the model's uptake by all relevant developers and managers. Note that the effort estimates herein only relate to the types of applications and projects that were used as basis for building the model.
- As developers and managers are using the model, it is important to obtain feedback on its use, as it may need to be updated at some point. Those who have participated in building the model should ideally be the ones engaged in any model updates that take place.

Discussions with the Development Team(s)

Even if the company is small and most estimates are prepared by the company owner and/or project manager, we suggest that developers be presented with the model, its value, its detailed description and nomenclature, so a common vocabulary can be used between developers and manager whenever discussions relating to effort estimation take place.

Decision Making Between Project Managers

Project managers can use the model for decision making, where such discussions can involve the more experienced managers who participated in the model building process, and also more junior managers, who can not only participate in the decisions and discussions, but who can also learn via an internalisation process (explicit model driving changes to their tacit knowledge).

Checks and Balances for Effort Estimates Provided by Contractors

The model can be used to check whether the effort estimates suggested by contractors working on parts of a project are reasonable or not. This can be a very useful approach as the model can even be presented to the contractors as part of the discussion relating to their estimates (in case they have overestimated their estimates a great deal).

Discussions with Clients

The model can be shown to clients as a way to provide them reassurance that the effort estimates being put forward are not simply educated guesses. The effort ranges associated with the highest probability for the development effort factor can also be used in order to discuss different costs and also durations for delivering the application.

Meetings with Clients

Project managers and/or requirements analysts can take the model to requirements elicitation meetings and use it as a guide in order to obtain some of the evidence to be entered in the model, so to get an effort estimate. Such an approach can be effective and help make the elicitation meetings focused, in particular whenever clients want quick cost estimates based on very short elicitation meetings.

Seminar to Other Branches and/or Events on Best Practices

The model can be shown to other departments, divisions or branches within the same company as part of a wider strategy to use such a modelling approach to improve effort estimates, and other aspects too (e.g., quality prediction). In addition, such a model, or experiences from using it, can also be presented at industry events as examples of best practice.

Conclusion

This chapter has presented a few suggestions on how Bayesian network models can be used by companies to improve their effort estimation processes. All the suggestions given are based on post-mortem meetings with project managers from companies with which we collaborated in building effort estimation Bayesian network models.

Conclusions

14

Introduction

A cornerstone of Web project management is effort estimation, the process by which effort is forecasted and used as basis to predict costs and allocate resources effectively, so enabling projects to be delivered on time and within budget. Effort estimation is a very complex domain where the relationship between factors is nondeterministic and has an inherently uncertain nature. For example, assuming there is a relationship between development effort and an application's size (e.g., number of Web pages, functionality), it is not necessarily true that increased effort will lead to larger size. However, as effort increases so does the probability of larger size. Effort estimation is a complex domain where corresponding decisions and predictions require reasoning with uncertainty.

Within the context of Web effort estimation, numerous studies investigated the use of effort prediction techniques. However, to date, only Mendes [1–6] investigated the explicit inclusion and use of uncertainty, inherent to effort estimation, into models for Web effort estimation. Mendes [1–3] built a hybrid Bayesian network (BN) model (structure expert-driven and probabilities data-driven), which presented significantly superior predictions than the mean- and median-based effort [2], multivariate regression [1–3], case-based reasoning and classification, and regression trees [3]. Mendes [4], and Mendes and Mosley [6] extended their previous work by building respectively four and eight BN models (combinations of hybrid and data-driven). These models were not optimised, as previously done in Mendes [1–3], which might have been the reason why they presented significantly worse accuracy than regression-based models. Finally, Mendes et al. [7], and Mendes [5, 8, 9] describe case studies where an expert-based Web effort estimation BN model was successfully built and used to estimate effort for projects developed by Web companies in Auckland, New Zealand. This chapter combines the experience and findings resulting from these four case studies, plus another two (yet to be published), revisits the process employed to build and validate the BN models, and discusses lessons learned.

E. Mendes, *Practitioner's Knowledge Representation*, DOI 10.1007/978-3-642-54157-5_14,   193
© Springer-Verlag Berlin Heidelberg 2014

Fig. 14.1 Example of a BN
model and two CPTs

CPT for node size (new Web pages)	
Low	0.2
Medium	0.3
High	0.5

CPT for node total effort (TE)			
Size (new Web pages)	*Low*	*Medium*	*High*
Low	0.8	0.2	0.1
Medium	0.1	0.6	0.2
High	0.1	0.2	0.7

As presented in Chap. 6, a BN is a model that supports reasoning with uncertainty due to the way in which it incorporates existing complex domain knowledge [10]. It represents knowledge using two parts. The first, the qualitative part, represents the structure of a BN as depicted by a directed acyclic graph (digraph) (Fig. 14.1). The digraph's nodes represent the relevant variables (factors) in the domain being modelled, which can be of different types (e.g., observable or latent, categorical). The digraph's arcs represent the causal relationships between variables, where relationships are quantified probabilistically. The second, the quantitative part, associates a node conditional probability table (CPT) to each node, its probability distribution. A parent node's CPT describes the relative probability of each state (value); a child node's CPT describes the relative probability of each state conditional on every combination of states of its parents (e.g., in Fig. 14.1, the relative probability of total effort (TE) being "low" conditional on size (new Web pages) (SNWP) being "low" is 0.8). Each column in a CPT represents a conditional probability distribution and therefore its values sum up to 1 (or 100, depending on how this is set when entering the probabilities in the CPTs) [10]. Once a BN is specified, evidence (e.g., values) can be entered into any node, and probabilities for the remaining nodes are automatically calculated using Bayes' rule [11]. Therefore BNs can be used for different types of reasoning, such as predictive and "what-if" analyses to investigate the impact that changes on some nodes have on others [12].

Within the context of Web effort estimation there are issues with building data-driven or hybrid Bayesian models, as follows:

1. Any dataset used to build a BN model should be large enough to provide sufficient data capturing all (or most) relevant combinations of states amongst variables such that probabilities can be learnt from data, rather than elicited manually. Under such circumstances, it is very unlikely that the dataset would contain project data volunteered by only a single company (single-company dataset). As far as we know, the largest dataset of Web projects available is the Tukutuku dataset (195 projects) [13]. This dataset has been used to build data-driven and hybrid BN models; however, results have not been encouraging overall, and we believe one of the reasons is due to the small size of this dataset.

2. Even when a large dataset is available, the next issue relates to the set of variables part of the dataset. It is unlikely that the variables identified represent all the factors within a given domain (e.g., Web effort estimation) that are important for companies that are to use the data-driven or hybrid model created using this dataset. This was the case with the Tukutuku dataset, even though the selection of which variables to use had been informed by two surveys [13]. However, one could argue that if the model being created is hybrid, then new variables (factors) can be added to, and existing variables can be removed from the model. The problem is that every new variable added to the model represents a set of probabilities that need to be elicited from scratch, which may be a hugely time consuming task.

3. Different structure and probability learning algorithms can lead to different prediction accuracy [6]; therefore one may need to use different models and compare their accuracy, which may also be a very time consuming task.

4. When using a hybrid model, the BN's structure should ideally be jointly elicited by more than one domain expert, preferably from more than one company; otherwise the model built may not be general enough to cater for a wide range of companies [6]. There are situations, however, where it is not feasible to have several experts from different companies cooperatively working on a single BN structure. One such situation is when the companies involved are all consulting companies potentially sharing the same market. This was the case within the context of this research.

5. Ideally the probabilities used by the data-driven or hybrid models should be revisited by at least one domain expert, once they have been automatically learned using the learning algorithms available in BN tools. However, depending on the complexity of the BN model, this may represent having to check thousands of probabilities, which may not be feasible. One way to alleviate this problem is to add additional factors to the BN model in order to reduce the number of causal relationships reaching child nodes; however, all probabilities for the additional factors would still need to be elicited from domain experts.

6. The choice of variable discretisation, structure learning algorithms, parameter estimation algorithms and the number of categories used in the discretisation all affect the accuracy of the results, and there are no clear-cut guidelines on what would be the best choice to employ. It may simply be dependent on the dataset being used, the amount of data available, and trial and error to find the best solution [6].

Therefore, given the abovementioned constraints, as part of two government-funded projects on using Bayesian networks to Web effort estimation (New Zealand and Brazilian governments), several expert-based company-specific Web effort BN models were built and validated, with the participation of five Web companies in New Zealand, and one company in Brazil. The development and successful deployment of these six models is the subject and contribution of this book.

Note that we are not suggesting that data-driven and hybrid BN models should not be used. On the contrary, they have been successfully employed in numerous domains [14]; however, the specific domain context of this paper—that of Web effort estimation, provides other challenges (described above) that lead to the development of solely expert-driven BN models.

We would also like to point out that we have explicitly emphasised that the models focus on Web effort estimation because, in our view, Web and software development differ in a number of areas, such as: application characteristics, primary technologies used, approach to quality delivered, development process drivers, availability of the application, customers (stakeholders), update rate (maintenance cycles), people involved in development, architecture and network, disciplines involved, legal, social, and ethical issues, and information structuring and design. A detailed discussion on this issue is provided in [15]. However, despite the differences between Web and software development, we believe that our contribution goes beyond the area of Web engineering, given that the process presented herein can also be used to build BN models for any IT company that estimates development effort for their projects.

General Process Employed to Build BNs

As detailed in Chap. 6, the BNs that are the focus of this book were built and validated using an adaptation of the expert-based knowledge engineering of Bayesian networks (EKEBN) process proposed in [14] (Fig. 14.2). Within the context of this work the author was the knowledge engineer (KE), and Web project managers from several well-established Web companies in either New Zealand or Brazil were the domain experts (DEs).

The three main steps within the adapted KEBN process are the structure building, uncertainty quantification, and model validation. This process iterates over these steps until a complete BN is built and validated. Each of these three steps is detailed next, and is presented in Fig. 14.2.

Structure Building This step represents the qualitative component of a BN, which results in a graphical structure comprised of, in our case, the factors (nodes, variables) and causal relationships identified as fundamental for effort estimation of Web projects. In addition to identifying variables and causal relationships, this step also comprises the identification of the states (values) that each variable should take, and if they are discrete or continuous. In practice, currently available BN tools require that continuous variables be discretised by converting them into multinomial variables, which is also the case with the BN software used in this study. The BN's structure is refined through an iterative process. This structure construction process has been validated in previous studies [12, 14, 16–18] and uses the principles of problem solving employed in data modelling and software development [19]. As will be detailed later in this chapter, existing literature in Web effort estimation, and knowledge from the DEs were employed to elicit the Web effort

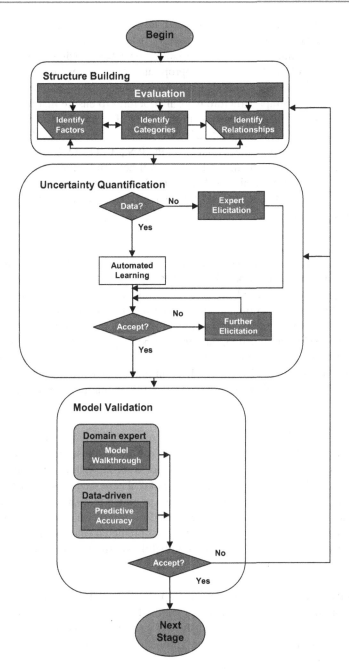

Fig. 14.2 EKEBN process

BNs' structures. Throughout this step the KE also evaluates the structure of the BN, done in two stages. The first entails checking whether variables and their values have a clear meaning, all relevant variables have been included, variables are named conveniently, all states are appropriate (exhaustive and exclusive) and includes a check for any states that can be combined. The second stage entails reviewing the BN's graph structure (causal structure) to ensure that any identified d-separation dependencies comply with the types of variables used and causality assumptions. D-separation dependencies are used to identify variables influenced by evidence coming from other variables in the BN [10, 11]. Once the BN structure is assumed to be close to final, the KE may still need to optimise this structure to reduce the number of probabilities that need to be elicited or learnt for the network. If optimisation is needed, techniques that change the causal structure (e.g., divorcing [10]) are employed.

Uncertainty Quantification This step represents the quantitative component of a BN, where conditional probabilities corresponding to the quantification of the relationships between variables [10, 11] are obtained. Such probabilities can be attained via expert elicitation, automatically from data, from existing literature, or using a combination of these. When probabilities are elicited from scratch, or even if they only need to be revisited, this step can be very time consuming. In order to minimise the number of probabilities to be elicited, some techniques have been proposed in the literature [16, 20, 21]. In addition, we have also recently proposed a technique to reduce the time needed for probability elicitation, to be discussed later.

Model Validation This step validates the BN resulting from the two previous steps, and determines whether it is necessary to revisit any of those steps. Two different validation methods are generally used—model walk-through and predictive accuracy. Model walk-through represents the use of real case scenarios that are prepared and used by DEs to assess if the predictions provided by a BN correspond to the predictions experts would have chosen based on their own expertise. Success is measured as the frequency with which the BN's predicted value for a target variable (e.g., quality, effort) that has the highest probability corresponds to the experts' own assessment.

Predictive accuracy uses past data (e.g., past project data), rather than scenarios, to obtain predictions. Data (evidence) is entered on the BN model, and success is measured as the frequency with which the BN's predicted value for a target variable (e.g., quality, effort) showing the highest probability corresponds to the actual value from past data.

Process Used to Build the Expert-Based BNs

This subsection revisits the adapted EKEBN process (Fig. 14.2), detailing the tasks carried out for each of the three main steps that form part of that process. Before starting the elicitation of the Web effort BN models, the domain experts (DEs) from all participating Web companies were presented with an overview of Bayesian network models, and examples of "what-if" scenarios, using a made-up BN. This, we believe, facilitated the entire process as the use of an example, and the brief explanation of each of the steps in the KEBN process, provided a concrete under-standing of what to expect. We also made it clear that the KE was a facilitator of the process, and that the Web companies' commitment was paramount for the success of the collaboration. The effort required by each company to have their BN models created and the characteristics of each model are detailed in Table 14.1.

The DEs who took part in the case studies were all project managers of well-established Web companies in either Auckland (New Zealand), or Rio de Janeiro (Brazil), each with at least 10 years of experience in project management. These companies varied in their size, measured as the total number of employees. In addition, all six companies were consulting companies and as such, developed a wide range of Web applications, from static and multimedia-like to very large e-commerce solutions. All six companies employed a wide range of Web techno-logies, thus also enabling the development of Web 2.0 and Web 3.0 applications. Finally, when approached, they were all looking at improving their current effort estimates, and agreed to participate for two main reasons: (1) because the models being created were single-company models geared towards their specific needs; (2) and also because their expertise and participation were acknowledged as essential to eliciting the models.

Detailed Structural Development and Parameter Estimation In order to identify the fundamental factors that the DEs took into account when preparing a project quote we used the set of variables from the Tukutuku dataset [13] as a starting point (Table 14.2). We first sketched them out on a whiteboard, each one inside an oval shape, and then explained what each one meant within the context of the Tukutuku project. Our previous experience eliciting BNs in other domains (e.g., ecology, resource estimation) suggested that it was best to start with a few factors (even if they were not to be reused by the DE), rather than to use a "blank canvas" as a starting point [7].

Within the context of the Tukutuku project, based on collected data, a new high-effort feature/function and a high-effort adapted feature/function require respec-tively at least 15 and 4 h to be developed by one experienced developer.

Once the Tukutuku variables had been sketched out and explained, the next step was to remove all variables that were not relevant for the DEs, followed by adding to the white board any additional variables (factors) suggested by them. This entire process was documented using digital voice recorders and also text editors. We also documented descriptions and rationale for each factor proposed by the DEs.

Table 14.1 Characteristics of the Bayesian network models and number of DEs

| Characteristics | Companies (country: New Zealand (NZ) or Brazil (BR)) | | | | | |
	A (NZ)	B (NZ)	C (NZ)	D (NZ)	E (NZ)	F (BR)
Number of DEs	1	1	2	2	7/2	1
Number of employees	~5	~5	~20	~30	~100	~30
Number of 3-h elicitation sessions	12	6	8	12	12/12	20
Total hours to elicit and validate model	36	18	24	36	98	60
Effort to elicit and validate model (person-hours)	72	36	72	108	324	120
Number of factors	14	13	34	33	38	19
Number of relationships	18	12	41	60	50	37
Number of past projects used as validation set	22	8	11	22	22	9

Table 14.2 Tukutuku variables

	Variable name	Description
Project data	*TypeProj*	Type of project (new or enhancement)
	nLang	Number of different development languages used
	DocProc	If project followed defined and documented process
	ProImpr	If project team involved in a process improvement programme
	Metrics	If project team part of a software metrics programme
	DevTeam	Size of a project's development team
	TeamExp	Average team experience with the development language (s) employed
Web application	*TotWP*	Total number of Web pages (new and reused)
	NewWP	Total number of new Web pages
	TotImg	Total number of images (new and reused)
	NewImg	Total number of new images created
	Num_Fots	Number of features reused without any adaptation
	HFotsA	Number of reused high-effort features/functions adapted
	Hnew	Number of new high-effort features/functions
	TotHigh	Total number of high-effort features/functions
	Num_FotsA	Number of reused low-effort features adapted
	New	Number of new low-effort features/functions
	TotNHigh	Total number of low-effort features/functions

The factors proposed were indeed influenced by DEs' hunches and insights; however, DEs' decisions and choices were also very much influenced by their solid previous experience managing Web projects, and estimating development effort.

Next, we identified the possible states that each factor would take. All states were discrete. Whenever a factor represented a measure of effort (e.g., total effort), we also documented the effort range corresponding to each state, to avoid any

Fig. 14.3 An example of a
cause and effect relationship

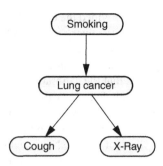

future ambiguity. For example, to one of the participating Web companies, "very low" total effort corresponded to 4+ to 10 person-hours, etc. Once all states were identified and thoroughly documented, it was time to elicit the cause and effect relationships. As a starting point to this task we used a simple medical example from [14] (Fig. 14.3).

This example clearly introduces one of the most important points to consider when identifying cause and effect relationships—the timeline of events. If smoking is to be a cause of lung cancer, it is important that the cause precedes the effect. This may sound obvious with regard to the example used; however, it is our view that the use of this simple example significantly helped the DEs understand the notion of cause and effect, and how this related to Web effort estimation and the BNs being elicited. Once the cause and effect relationships were identified, we worked on the elicitation of probabilities to quantify each of the cause and effect relationships previously identified. In all four cases, there was an iterative process between the structural development and parameter elicitation steps.

Detailed Model Validation Both model walkthrough and predictive accuracy were used to validate all six Web effort BN models, where the former was the first type of validation to be employed in all cases. DEs used different scenarios to check whether the node total_effort would provide the highest probability to the effort state that corresponded to the DE's own suggestion. However, it was also necessary to use data from past projects, for which total effort was known, in order to check the model's calibration. Table 14.1 details the number of projects used by each company as validation set. In all cases, DEs were asked to use as validation set a range of projects presenting different sizes and levels of complexity, and being representative of the types of projects developed by their Web company.

For each project in a validation set, evidence was entered in the BN model, and the effort range corresponding to the highest probability provided for "total effort" was compared to that project's actual effort. Whenever actual effort did not fall within the effort range associated with the category with the highest probability, there was a mismatch; this meant that some probabilities needed to be adjusted. In order to know which nodes to target first we used a sensitivity analysis report, which provided the effect of each parent node upon a given query node. Within our context, the query node was "total effort".

Whenever probabilities were adjusted, we re-entered the evidence for each of the projects in the validation set that had already been used in the validation step to ensure that the calibration already carried out had not been affected. This was done to ensure that each calibration would always be an improvement upon the previous one. Once all projects were used to calibrate a model, the DE(s) assumed that the validation step was complete.

Each of the five New Zealand BNs has been in production for at least 18 months, and the Brazilian BN has been in production since May 2011.

Common Patterns

All six models were BNs targeting at the estimation of effort for new projects; therefore, we also looked at combining part of results from the six different case studies focusing at two specific points: (1) to identify the total set of factors selected by the Web companies; (2) to identify the amount of overlap between companies, measured using a vote counting approach. In order to identify the total set of factors using a common terminology, we employed a methodology that we have previously proposed, which uses the contextual meaning of each factor in order to match factors across different BN models. The details relating to this methodology are outside the scope of this chapter and book; however, for those interested, they are documented in [22].

Apart from total effort, which was identified by all participating companies, there were three factors that were chosen by five of the six Web companies:

- Average project team experience with technology
- Effort to program features
- Project management effort

All five BN models showed these factors affecting total effort. This includes project management, shown to be affected by other factors such as the number of features to develop, or project risk.

The next set of factors selected by four Web companies were the following:

- Adaptation effort of features off the shelf
- Development effort of new features
- Effort to develop user interface
- Project risk factor
- Effort production testing (Table 14.3)

Except for project risk factor and average project team experience with technology, all the remaining factors related to the effort to accomplish certain tasks, such as adapting or developing a new feature, testing and interface design. Note that these factors are very much related to more dynamic Web applications, which offer

Table 14.3 Common factors amongst companies

Factor	Scale	No. of companies
Total effort	Person-hours	6
Average project team experience with technology	Years	5
Effort to program features	Person-hours	5
Project management effort	Person-hours	5
Adaptation effort of features off the shelf	Person-hours	4
Development effort of new features	Person-hours	4
Effort to develop user interface	Person-hours	4
Project risk factor	UD (e.g., low, medium, high)	4
Effort production testing	Person-hours	4
Client personality difficulty	UD (e.g., low, medium, high; good, normal, bad)	3
Effort producing animations using software	Person-hours	3
Effort to implement the Web application	Person-hours	3
Effort to produce requirements documentation	Person-hours	3
Effort to produce template mock-up	Person-hours	3
Effort to produce Web pages	Person-hours	3
How much technical planning	UD (e.g., low, normal, high)	3
Number of features off the shelf	Integer	3
Number of features off the shelf adapted	Integer	3
Number of new Web pages	Integer	3
Effort post-release testing	Person-hours	3
Deployment time	UD (e.g., short, normal)	3
Development team size	Integer	3
Effort images manipulation	Person-hours	3
Effort to integrate new and reused features	Person-hours	3
Quality of project management	UD (e.g., abysmal, low, normal, high)	3
Technical planning effort	Person-hours	3
Level of integration between features	UD (e.g., low, medium, high)	3
Effort to design content	Person-hours	3
Effort programming animations	Person-hours	2
Effort template look and feel	Person-hours	2
Web company's hosting control	UD (e.g., client in-house, shared, dedicated, in-house)	2
Is development process documented?	Yes/no	2

(continued)

Table 14.3 (continued)

Factor	Scale	No. of companies
Number of features requiring high effort to create	Integer	2
Number of features requiring low effort to create	Integer	2
Number of features requiring medium effort to create	Integer	2
Number of key client's people	Integer	2
Number of reused Web pages	Integer	2
Number of third parties involved	Integer (e.g., subcontractors, printing, SMS gateways, hosting providers, domain registration, payment providers)	2
Quality of in-house existing code	UD (e.g., low, normal, high)	2
Quality of third-party deliverables	UD (e.g., low, high)	2
Type of project	UD (e.g., new, enhancement)	2
Number of natural languages used	Integer	2
Total third-party inexperience	UD (e.g., low, medium, high)	2
Unknown technology risk	UD (boolean)	2
Amount of text per application	UD (e.g., low, medium, High)	1
Client application domain literacy	UD (e.g., low, medium, High)	1
Client's existing online presence	UD (e.g., small, extensive, none)	1
Development process model	UD (e.g., conventional, waterfall, extreme)	1
Number of images requiring high effort to manipulate	Integer	1
Number of images requiring low effort to manipulate	Integer	1
Number of images requiring medium effort to manipulate	Integer	1
Number of Web page templates	Integer	1
Level of usability	UD (e.g., low, medium, high)	1
Similarity to previous projects	UD (similarity of domain/functionality/design; e.g., low, medium, high)	1
Legacy browser support	UD (e.g., yes, no)	1
Effort to implement accessibility	Person-hours	1
Forum feature	UD (Boolean)	1
User sign-up feature	UD (Boolean)	1
Auction system feature	UD (Boolean)	1
Types of listing features	UD categories	1

(continued)

Table 14.3 (continued)

Factor	Scale	No. of companies
Gallery feature (number of controls)	UD (number of widgets)	1
Shopping cart feature	UD (Boolean)	1
Event calendar feature	UD (Boolean)	1
Number of blogs	Integer	1
Number of poll	Integer	1
Mailing list feature	UD (Boolean)	1
Effort to produce user documentation	Person-hours	1
Tight schedule	Boolean	1
Template design uniqueness	UD (e.g., template standard, template high, custom-medium, custom-high	1
Effort to implement the template	Person-hours	1

a large set of features (this requiring more detailed testing). It is also interesting that the effort to develop the user interface was also chosen by four of the six companies. Nowadays, with the plethora of Web technologies and possibilities available, good interface design and usability can also add very much so to a company's competitive advantage on the global market.

All six BNs presented several effort-related factors as predictors of total effort. We believe that this occurred because all models also reflected the specific effort estimation workflows employed by each of the participating companies.

Lessons Learned

The work that was detailed in Chaps. 7–12 has provided numerous lessons, as follows:

First: Engaging with industry. At the start of this research, in order to reach out to industry, we invited the local NZ IT industry to attend a seminar about Web effort estimation and how to improve their estimates. The seminar provided an introduction to using expert-based BNs, their value as estimation tools and their capability for running "what-if" scenarios. Many of the participating companies saw the immediate value in such an approach, in particular because it enabled the very close and fundamental participation of in-house domain experts while building and validating the company-specific model. Several companies signed up to collaborate.

Second: Time constraints. Depending on the complexity of the BN model, the elicitation of probabilities can be very time consuming and last several months. At the start of the research project we did not have the means to provide the automatic generation of probabilities, so all probabilities had to be elicited manually based on expert knowledge. This was a drawback, given that the time needed to develop a full-fledged model became prohibitive to some companies. As a consequence, out of the initial set of ten NZ participating companies, only five remained until the full completion of their BN models. Motivated by this issue, we also investigated mechanisms to enable the automatic generation of probability tables. A solution was devised and used with two NZ companies that participated more recently in this research. This solution comprised the comparison between different probability generation algorithms and expert-driven probabilities. Therefore we needed first to have completed the probability elicitation and validation of several BN models so to have a basis for comparison with the proposed algorithms. A tool was implemented as a result of this work; further details can be found in [23, 24].

Third: Value for a company. Except for the company in Brazil, the other participating companies were contacted for post-mortem interviews. The main points highlighted were the following:

- The elicitation process enabled experts to think deeply about their effort estimation process and the factors taken into account during that process, which in itself was considered advantageous to the companies. This has been pointed out to us by all the DEs interviewed.
- Once a BN model was validated, DEs started to use their model not only for obtaining better estimates than the ones previously prepared by subjective means, but also sometimes as means to guide their requirements elicitation meetings with prospective clients. They targeted their questions at obtaining evidence to be entered in the model as the requirements meetings took place; by doing so they basically had effort estimates that were practically ready to use for costing the projects, even when meetings with clients had short durations. Such change in approach proved to be extremely beneficial to the companies given that all estimates provided using the models turned out to be more accurate on average than the ones previously obtained by subjective means.
- Clients were not presented the models due to their complexity; however, by entering evidence while a requirements elicitation meeting took place the DEs were able to optimize their elicitation process by being focused and factor-driven.
- One of the participating companies, the largest company in total number of employees, and also the one that built the largest BN model, provided the following feedback: The DEs who participated in the causal structure and probabilities' elicitation completely changed their approach to estimating effort. These DEs presented the BN model to all of their development teams, and asked that from that point onwards every estimate for any task should be based on the factors that had been elicited. This means that an entire team started to use the factors that have been elicited, as well as the BN model, as basis for their effort and risk-estimation sessions. In addition, the DEs presented the model at a meeting with other company branches, so to detail how the Auckland branch was estimating effort and risk for their healthcare projects. The other branches were so impressed, in particular the one from the US, that they increased the number of healthcare software projects outsourced to the NZ branch, as they recognised the benefits of using a model that represented factors and uncertainties. Overall, such change in approach provided extremely beneficial to the company.

All the companies remained positive and very satisfied with the results. We believe that the successful development of these six Web effort BN models was greatly influenced by the commitment of the participating companies, and also by the DEs' experience estimating effort.

References

1. Mendes E (2007) Predicting web development effort using a Bayesian network. In: Proceedings of EASE'07, Swinton, UK, pp 83–93
2. Mendes E (2007) The use of a Bayesian network for web effort estimation. In: Proceedings of international conference on web engineering, LNCS 4607, pp 90–104
3. Mendes E (2007) A comparison of techniques for web effort estimation. In: Proceedings of the ACM/IEEE international symposium on empirical software engineering, Madrid, pp 334–343
4. Mendes E (2008) The use of Bayesian networks for web effort estimation: further investigation. In: Proceedings of ICWE'08, Yorktown Heights, NJ, pp 203–216
5. Mendes E (2011) Building a web effort estimation model through knowledge elicitation. In: Proceedings of the 13th international conference on enterprise information systems, pp 128–135
6. Mendes E, Mosley N (2008) Bayesian network models for web effort prediction: a comparative study. Trans Softw Eng 34(6):723–737
7. Mendes E, Pollino C, Mosley N (2009) Building an expert-based web effort estimation model using Bayesian networks. In: Proceedings of the EASE conference, Swinton, UK, pp 1–10
8. Mendes E (2011) Improving project management of healthcare projects through knowledge elicitation. In: Miranda IM, Cruz-Cunha MM (eds) Handbook of research on ICTs for healthcare and social services: developments and applications. IGI Global, Hershey
9. Mendes E (2011) Uncertainty-based software effort estimation. The International Function Points Users Group (IFPUG) Guide to IT and Software Measurement, CRC press
10. Jensen FV (1996) An introduction to Bayesian networks. UCL Press, London
11. Pearl J (1988) Probabilistic reasoning in intelligent systems. Morgan Kaufmann, San Mateo, CA
12. Fenton N, Marsh W, Neil M, Cates P, Forey S, Tailor M (2004) Making resource decisions for software projects. In: Proceedings of ICSE'04, pp 397–406
13. Mendes E, Mosley N, Counsell S (2005) Investigating web size metrics for early web cost estimation. J Syst Softw 77(2):157–172
14. Woodberry O, Nicholson A, Korb K, Pollino C (2004) Parameterising Bayesian networks. In: Proceedings of Australian conference on artificial intelligence, Berlin, pp 1101–1107
15. Mendes E, Mosley N, Counsell S (2005) The need for web engineering: an introduction. In: Mendes E, Mosley N (eds) Web engineering. Springer, Berlin, pp 1–28. ISBN 3-540-28196-7
16. Druzdzel MJ, van der Gaag LC (2000) Building probabilistic networks: where do the numbers come from? IEEE Trans Knowl Data Eng 12(4):481–486
17. Mahoney SM, Laskey KB (1996) Network engineering for complex belief networks. In: Proceedings of 12th annual conference on uncertainty in artificial intelligence, San Francisco, CA, pp 389–396
18. Neil M, Fenton N, Nielsen L (2000) Building large-scale Bayesian networks. Knowl Eng Rev 15(3):257–284
19. Studer R, Benjamins VR, Fensel D (1998) Knowledge engineering: principles and methods. Data Knowl Eng 25:161–197
20. Das B (2004) Generating conditional probabilities for Bayesian networks: easing the knowledge acquisition problem. http://arxiv.org/pdf/cs/0411034v1.pdf. Accessed on 2008
21. Tang Z, McCabe B (2007) Developing complete conditional probability tables from fractional data for Bayesian belief networks. J Comput Civ Eng 21(4):265–276

22. Baker S, Mendes E (2010) Aggregating expert-driven causal maps for web effort estimation. In: Proceedings of the international conference on advanced software engineering & its applications (ASEA), vol 117, pp 264–282
23. Baker S, Mendes E (2010) Assessing the weighted sum algorithm for automatic generation of probabilities in Bayesian networks. In: Proceedings of the international conference on information and automation 2010 (ICIA 2010), Harbin, pp 867–873. doi: 10.1109/ICINFA.2010. 5512447
24. Baker S, Mendes E (2010) Evaluating the weighted sum algorithm for estimating conditional probabilities in Bayesian networks. In: Proceedings of the software engineering and knowledge engineering conference (SEKE 2010), pp 319–324

Index

A
Academic datasets, 59
Adaptation rules, 43–44
Advanced COCOMO, 34
Algorithmic techniques, 32
Amount of overlap between companies, 202
Analogy adaptation, 42
Artificial intelligence techniques, 38

B
Bayesian models, 194
Bayesian network (BN), 62–68
  model, 73
Bayes' theorem, 68

C
Calibration, 98, 104
Case-based reasoning (CBR), 38–39
Causal relationships, 62
Checks and balances for effort estimates by
  contractors, 191
Classification and Regression Trees (CART),
  44–47
COCOMO, 32–38
Combination of knowledge, 79–80
Common patterns, 202–205
Common vocabulary, 190
Concrete and industry-informed choices, 189
Conditional probability table (CPT), 62
Cost estimate, 29
Cost factors, 28

D
Data and/or knowledge on past finished
  projects, 28
Decision making, 190
  under uncertainty, 62

Discussions with clients, 191
Domain experts, 75
Duration estimate, 29
Dynamic Web applications effort estimation
  BN model, 141, 143

E
Effort estimate(ion), 28, 55, 193
  overview of process, 27–29
  purpose of, 27
EKEBN. *See* Expert-based knowledge
  engineering of Bayesian networks
  (EKEBN)
Elicitation exercise, 85
Engaging with industry, 205
Estimated size, 28
Estimation techniques, 59
Evidence, 66, 77, 100, 101
Expert-based Bayesian network models, 74, 76
Expert-based effort estimation, 29–32
Expert-based estimation drawbacks, 30
Expert-based knowledge engineering of
  Bayesian networks (EKEBN), 123, 196
  process, 75, 107, 199
  steps, 80
Explicitation of tacit knowledge, 75
Externalisation of knowledge, 78–79

F
Feature subset selection, 39–40

H
Healthcare effort and risk-prediction BN
  model, 107
Healthcare software effort and risk estimation
  BN model, 110

E. Mendes, *Practitioner's Knowledge Representation*, DOI 10.1007/978-3-642-54157-5,
© Springer-Verlag Berlin Heidelberg 2014

Printed in the United States
By Bookmasters